The Zetas describe themselves:

We are more an advanced version of the Ebens. We are not like what some describe as the Verdants. Biologically, we are cousins, but socially and culturally, we are separate, both in a moral context as well as a social context. We understand and respect life forms, whether more advanced or less advanced than us. We are more traveled and have acquired this attitude from our many contacts with many species.

ALIEN GAME CHANGER

Humans and Aliens Communicate

Mary Barr, Therapist
and J. Steven Reichmuth,
Telepathic Conduit

λ

Standing Wave Press

Printed in the United States of America.
Cover x-ray image by NASA/CXC/SSC
Cover design by Mary Barr
ISBN: 978-0-9895231-2-7

First Edition

First Printing, 2019

Non-fiction

Extraterrestrial beings
Human-alien encounters
Physics of the Cosmos

To those who are ready to study the cosmos and a non-Earth species.

Table of Contents

FOREWORD

The Zetas share their goals and life philosophy with us:

Life and happiness: to exist at one with the Universe and to cohabitate with other species in harmony and to pursue the unknown with all of our mental, physical and spiritual energy. It is with knowledge that those understandings make our coexistence with the Universe more intimate.

> Han, Zestra and Gen, physical beings from
> Zeta Reticuli I, Reticulum Constellation

ACKNOWLEDGEMENTS

Zetas explain their contributions to this book:

We know by your culture, that our names could not be in the book as credit, but there are many authors to this book. You and Steve will be the primary contributors, with you doing the writing. You will be our ambassadors. The contributions came from many of our species.

Han, Zestra and Gen

Chapter One

ALIEN GAME CHANGER
Humans and Aliens
Communicate

Mary Barr and Steve Reichmuth

This book is a continuation from Aliens Answer and Aliens Answer II. The work commenced in 2011 and continues to this day. These interviews are verbatim transcripts taken from hypnotic sessions performed by Mary Barr with Steve Reichmuth serving as the channel through which the Zetas answered questions posed by this therapist. Steve was never told in advance about any questions that were to be asked at each session.

The Beings, Han, Zestra and Gen, tell us that they are from the Reticulum star system. At times, they have given physical evidence of their existence such as long-distance electrical effects, disrupting pain in both Steve and Mary, taking a biopsy from Dr. Gene Lipson, causing an orb to appear in Steve's home, telling details of case histories before this therapist actually interviewed the clients, and sharing some scientific knowledge currently unknown to our scientists.

Session Thirty
February 12, 2012

Steve is inducted using visual images of floating down a stream in a canoe.

Han, you are most cordially invited. Are you available? (Long pause.)

Hello. This is Han.

Thank you so much for coming to visit with us.

It is a pleasure. Steve's mind needs to be relaxed.

Would you help him adjust the sound of his microphone?

Okay. Does that sound better?

Yes. Thank you for helping Steve.

We had a very unusual church service today at a church that I sometimes attend. When I use the term 'stuffed animal' what comes to your mind?

These are representative dolls, which you use sometimes for companionship.

You could say that. Earth children are given imitation animals made to look like a dog, or cat, or bunny or whatever. Yes. They do provide companionship, but there is actually more going on in this relationship. People were asked to bring their favorite stuffed animal to church. These were adults. Adults do not normally keep these stuffed animals, and yet they brought them. There was a ceremony of blessing the stuffed animals with water, done humorously, but also done very lovingly.

The person giving the blessing is actually a therapist who uses stuffed animals and dolls to help people project their fears and desires into an inanimate object. It is a way of finding out what is going on, but it is also a way for human beings to get rid of fears. Also, human children are separated from their parents when they sleep, and it is difficult for them, so the stuffed animal is the substitute parent.

The dolls perform a special role. They act as a substitute. They act as a conduit with which to express emotions. They act as a doorway into the soul of the concerned, serving as an emotional security blanket. It serves many things. It is a way for transference of emotional insecurity or security.

Sometimes it stands in for the child. The child will 'lecture' the stuffed animal as though the child is taking on the parental role. There is a role reversal.

It serves a constructive purpose in human emotional development. There is more going on than is often seen on the surface.

It is also a dry run for the child in developing relationships. They can practice.

A toy simulator.

I just wanted to share that with you. It is just kind of quizzical on the surface, but it has a true function.

The service that you attended today illustrated many subtleties going on under the surface, showing a more meaningful purpose. That is one of the fascinating things that we find about your species.

It allows you to see how we adapt and fulfill our needs without telepathy. The needs don't go away. They have to be met, but it is just interesting to see the different forms that it takes until we do become telepathic.

Ironically, it conceals a beguiling intent to distract attention to learn the truth. Such is what happens with a non-telepathic species. Since we can see through this, it is what makes your species interesting. It is the acting roles you unconsciously perform day-to-day in interaction between your species.

When we feel a certain type of emotion, the after-effect is such that we feel like we have 'won' or that we have accomplished what we needed to accomplish. It is a kind of satisfaction that comes from evoking certain emotions.

You must convince both yourselves as well as others of your species. We, as sentient beings, share more and more with each other each day. I find profound the many things we have discussed in our sessions. Humans have earned more of my respect. There is more to you than even you can comprehend in yourselves.

Don't you think that would come from our souls? It is not that our bodies or minds are so great or so complicated, but there is a living energy that animates us.

This energy. Yes. We acknowledge it.

We all have that kind of soul that brings its questions and experiences, a soul wanting us to experience as much as possible and to appreciate it. As such, I find life to be sweet beyond measure.

Zeta Reticuli History

Does your home planet have a molten core?

Yes, it does. We have, at times, shifts in our continents, including Earthquakes to expand this process. To an extent, it can be minimized, but the forces are very great. Our technology and our structures are designed to flow with the geology. These energy releases provide give and take with the stresses that are released. It is a planet that is larger than Earth, but very similar in other ways. It is what you might call a 'super Earth' from your context and the information you have recently discovered. That is a new term often expressed by humans. We have oceans and continents. We have continents that shift.

Do you have electromagnetic poles?

Yes.

Do they stay more or less in the same regions as a 'true' north and south?

Yes. We also have electromagnetic storms on our star that create what you call auroras. The position of our planet, with its high iron content, makes a pronounced magnetic field. The solar energies that are released from our star react in a similar manner to your star. This is not uncommon. We do not have large magnetic storms, but they exist as part of the natural activity that occurs.

Are your north and south poles covered with snow and ice?

Yes. Our polar regions. This is a natural consequence. Our planet is slightly tilted. This creates a temperate climate with fewer extremes. We have noticed this on other planets. This creates climate that is conducive to life, one where there are no radical temperature changes. There are seasons. The Earth could move into a situation where there would be more radical ice ages and warmer temperatures that would be too warm for agriculture. These are concerns, but this is a bit off into the future, if it should occur. The Earth goes through cycles. Its poles and tilt are similar to our own planet. We have moderate temperatures. We have tropical regions near the equator and polar ice at the poles.

Do you have more land mass?

We have less land mass in proportion. Our world is about twice the size of your world in diameter. However, the land mass surface areas are about the same as Earth's land masses. This means that the oceans are more predominant in proportion to the land, but this has been a blessing, providing a richer atmosphere for protecting us from solar rays. A robust ozone layer exists. Our magnetic poles provide astonishing solar activity.

It must be beautiful.

Imagine an Earth, only larger with more resources. Our planet's population has been controlled to an extent where there are no stresses from overpopulation. The economic and social factors have been moderated where the stresses on the planet allow us to live in peace. We live in harmony with our planet. We are blessed with large oceans that provide the oxygen and nitrogen combination along with carbon dioxide, similar to your Earth. We have a larger volume to maintain this with a reduced population so that each is in balance.

I can well imagine having that extra space to breathe and nature having the space to fully express itself.

We have a larger 'backyard,' I would say. Yet we still have our underground dwellings and technology. I have spoken of this. It is just under the surface.

Since you have so much water surface, are you using any of the water surface area for habitation?

We use tidal forces to create energy. We use gravity to create hydroelectric energy. We use fusion energies and more advanced energies that you have yet to discover, which are compatible with a green type environment. (Question not really answered.)

Do you take boats out onto the water?

Yes. Over the surface. Our boats are a romantic notion. We still use them. We use other craft that can operate over the water or under the water, but we still use objects that use buoyancy to float on the surface.

Would you be kind enough to imprint one of those objects into Steve's mind, the kind that uses buoyancy?

They are surface craft. They are not dissimilar. They use physics similar (to your boats) but some are unusual. Some even use lift. Some are powered by forces that propel themselves, using energy that displaces the water to propel across the water. We still use wind. It is efficient. if you are in no hurry.

I am imagining a ship that has very little contact with the surface of the water.

The sails are one technology, and they are light and strong. The hulls are slim and elegant and have less surface area to prevent friction with the water. They also have blades or hydrofoils that will lift the hull off of the water to minimize hydrodynamic drag. These are ancient technologies. Not much has changed through all the centuries to our present. Our oceans are similar to yours, due to mineral content and density.

Do you ever have tidal waves?

Yes. We have tsunamis, to use your Japanese word. These are often due to tidal forces under the surface on planets with molten cores. We have had, at times, influences from space. Meteors can create such energies. These dangers have been abated since much of our land dwellings are underground. These areas can be at risk of flooding, but we address this by other means. We exploit these natural energies.

Thank you. I am getting a much clearer picture now.

Contact

Monday morning of this last week, I was typing up these notes from our session. While I was typing, I felt a moderate amount of pressure on my head on the right side.

It was only me.

Well, I felt it, being sensitive to those things.

I hope it did not alarm you.

Oh, no. I am not alarmed by these things at all.

I noticed that.

I am used to contact.

It was a visit.

It was very kind of you. I just wished to acknowledge it.

Sometimes I am with you, sometimes with Steve, but not formally, as we are now. I enjoy your company by projecting my thoughts, and the thoughts of others, like a friendly ghost.

I know there are many levels in the way you reach out and make contact.

I hope it was a gentle experience.

If it had not been, I would have sent a message. (Therapist laughs.)

Yes. It is also with respect that I know that there are times you need to be away. I respect that, too. It is not a constant thing. I, too, have times that I must be away, and, at times, each of us are away.

The Universe

Remember my question from last time? It was whether or not, to your knowledge, a society has ever chosen to transition en masse into a spiritual existence. We talked a bit about how some societies have purposefully made a partial transition, but we never really addressed whether an entire culture or planet chose to do this.

We have spoken about species or life forms that have elevated to a more advanced state that are intriguing to us. You once asked about what we find interesting. The concept of an entire population can be viewed from several perspectives. Are we released from our physical limitations? In a sense, each life form passes to that state. In a sense, the energy is passed on. Then at the same time, there could be a natural evolutionary path to another plane of existence. It seems that we all inevitably wish to follow to that eventual goal. I would imagine a vast population of energy that has coalesced together to form some greater purpose, individually or collectively. I envision all focused with one purpose in mind: to exist in that plane, together, in peace.

Thank you. Are you familiar with a galaxy known as M83?

M83 is the human reference. It is a galaxy dissimilar from our galaxy in which humans and our species exist. It is a galaxy many light years away by your measurement. What is it that you wish to know?

There is an indication that there may be a species there that we call the Verdants. Where is their source galaxy?

The Verdants are a lesser evolutionary form. Sometimes they have been known as the 'greys.' They are a species that is very arrogant in some ways. They are a pervasive species in that galaxy and wish to be pervasive in this galaxy, if they can.

Do they overpopulate themselves intentionally?

They do. I understand the species you are referencing.

Other

Do you know about our friend whom we call Doc? His name is Gene.

Yes.

He submitted questions for discussion today. May I present them?

Yes, I will answer them as best I can.

Gene: Will our economic climate improve or deteriorate after the next Presidential election?

It will continue to improve, regardless of whoever is elected.

Gene: Is there such a concept that past lives influence present lives in obvious ways, for instance, the carryover of fears/phobias or life choices?

Past fates (lives) are influences that can have unconscious effects on one's present life. Much is shed, released. I am happy to say that there is a sense of good as we improve. Many of the old pathways are replaced with good and improved ways. There is a mix of wishing to improve one's life and memories of bad things that tend to be forgotten. The good, beneficial memories that have a practical use are prioritized. Life wishes to become more efficient and less hindered, if possible. Lingering energies can restrain and hold us back. These need to be released and forgotten. As we take on new lives, new ways are found to focus on what is good and practical.

Gene: That is very good news.

Yes. We share the same. Life has a way of creating new challenges. Whenever possible, old, bad energies need to be left behind.

Gene: Yes. Thank you.

Gene: In the next 20 years, what can we hypothesize the average human life expectancy to be?

Life expectancy will continue to improve as knowledge increases in health and technology. The limiting factors will be economic, not due to any current condition. The limiting factor is in the economic systems not having the ability to create social patterns where the benefits can be distributed equally among its population. It is what you call socialism. Humans have natural instincts to create a moral equality where there will be equality of access to health and good nutrition. This should not be suppressed. There will be elite who think that they know best, but they will only create divisions, which will only stifle the eventual good that will reach everyone. Those who 'know best' will actually be in the way. Eventually, this will come to a head, which will need to be reconciled. Ultimately, this will be good news for everyone.

Gene: We already know about the human papilloma virus and the type of cancer it causes. Is cancer caused by a virus?

There are many causes of cancer. Some can be triggered by viruses. There are preexisting conditions that make this possible.

Gene: Do we not already have that capacity to have cancer?

Yes. It is an anomaly where triggers can be activated, and then the cells tend to expand rapidly. This is caused by environmental concerns and certain preexisting genetic conditions. It is one that is complex and has many forms and many triggers. It is like juggling many balls in the air. It is one that can be addressed. It can be conquered in time.

Gene: Does that including one of the triggers being bacteria?

Not directly. They can create conditions that affect the immune system, and the immune system will weaken and make itself more vulnerable to those conditions that foster cancer growth. I hope I have explained this in a meaningful way.

Gene: I certainly understand it.

Good.

Gene: Are dreams representative of events of the past or of things to come in the future?

Dreams often represent the past. They also represent fantasies, 'what if' scenarios of what might have occurred. The future can be plans of what can be. They can also be reflections of past thoughts forming dreams of what could happen. They represent what could have been and what could be at the same time.

Gene: Is it possible that we keep meeting the same souls over and over again (in physical form)? Does the same soul exist in many forms in different lives?

Rephrased: Is reincarnation true and, if so, do we kind of stay in a group so that when we return, we are really meeting the same old souls, but in new bodies and new lifetimes?

Often this is the case across the same time span and span of life. Life transforms. Souls continue, sometimes in 'flocks' gauged in natural physical life cycles. There can be the same life mates who will naturally associate with other souls based on whom we are repelled from or attracted to. This will continue to facilitate resolving unfinished business that most souls are unaware of consciously. In the human mind, I can see an analogy. The actor, Orson Wells, and his same actors on the radio who often would appear together with other actors. This association at the Mercury Theatre continued from radio to film with the same actors in other artistic media. This is an analogy for the soul encountering groups to which it is naturally attracted. For various reasons, energies and compatibilities of attractiveness seem to coalesce as the beings that they are, souls in purpose, intellect, and in emotion. You may refer to it by adapting your word phrase as 'spiritual chemistry.'

Gene: Do we contaminate our bodies with preservatives and manufactured medicines?

I am smiling. 'Doc' is the ice cream of life. His flavor is yet to be determined. This is said with humor and affection. I can see what concerns him. Sometimes these concerns are rather amusing in their confined perspectives. I wish I could help him let go of the things that close him off and release him to fly like the Peter Pan in your fables that he wishes to be. I think that the things that concern him should be released. In time, this will happen. He is a good soul and one that has done much human good. With his skill, love, and humanity, he has enabled others who have suffered to ease their troubles. He has brought so much life into the world in his role as a pediatrician. Not only is his intelligence bright, but his energy will glow brightly in many humans in the future. It is a bright star that he inhabits. He need have no worries concerning his worldly issues. Still, he holds onto them. It is his confounding affectionate Jewish questions that make me laugh with love that he does not need to worry about. It is hard to explain. He will happily be released from this. I hope I have expressed this appropriately. My comments are meant with the utmost respect and affection for him. He is less aware of us than we are of him.

Thank you. He is bright enough. He will get it. He has been a good contributor.

He will get it, and he will catch up with us. He is not far behind. At times, he is also, in a reflective awesome way, ahead of even me.

Will we (humans) eventually recognize that there are such things as past lives and life after death?

Yes. These are the things that we commonly talk about. This is what makes me laugh. Here they are. They are all around us (spirits). This is the very thing that we are speaking about.

I think, with him, that it depends upon who is telling him these things.

I, Han, am telling him this, with assurance. I touch his forehead in this. Yes. All of this is here, and he has so much joy to look forward to.

He wants to keep on helping children, of course.

And he will, with his transcending soul, not only help the children on Earth, but also the children on many worlds. His energies have so much to look forward to that he has not conceived of yet.

I am excited for him.

I am merely reflecting on all that he has to look forward to along with the rest of us. This is also true for Mary, Steve, and even Charlene. I see this little child that is released. It will give happiness and uncluttered freedom. As we let go, all souls will travel in the same direction.

Thank you for that.

I wanted to talk to you about what we discussed earlier, the great truths, and put ideas out there for discussion. We humans see truth a little differently. We have what we call personal truths, social truths, and universal truths. To give you an example, a common expression is: 'This, too, shall pass.' I would put that thought in the category of being a universal truth.

Do you see that phrase as a universal truth, or would you say it needs qualification?

I don't understand. I am sorry. Life? This, too, shall pass?

It means, in essence, everything changes.

Such is life.

For example, does an atom change its structure over time?

There is an ultimate life that transcends, and life transforms from one version to another to a higher self, in time. This is the only thing I can add.

Perhaps we could say, for example, 'All life transitions.' That might be a more accurate statement.

Yes. We are in our thought sharing the same meaning, both in the question and the answer. I am only struggling to express this in a language we share, too. All things transition and elevate us, too. It does not regress. This is one hope we all share. I am respectful of that.

Is there something that you might like to share with us that you would consider a universal truth?

Yes. I can share several with you. Mathematics is one universal truth. It is also a universal language. Another universal truth is physics, with which mathematics is closely allied. There is always an impulse for unification to its extreme that seeks a theory of everything. We seek one universal, single, generative equation for all we see and understand in nature. However, this becomes universally less clear whether such a theory would be a simplification, given the proliferation of dimensions and universes it might entail. Nonetheless, unification, in all its forms, remains a major preoccupation with all the sentient inhabitants of the Universe. The universal engine that propels this desire for unification comes simply when the highest compliment one can give another is to express that you made them think.

Open Session for Comments or Questions

Because Steve needs some rest, is there anything you wish to share before we allow him his rest?

He is starting to come back.

Are you with me, Han? Or is Steve coming back now.

I am here. Steve is somewhere distant. Steve feels lost.

Why don't we bring him in with us? Let him come in a little and exchange with us, so he does not feel so separated.

Steve, we are going to bring you in on the conversation so you feel more connected. Han is here. I am here. You are part of this. You may ask a question.

Steve: Thank you. Sometimes I feel left out. Other times, if included, I would feel strange and unneeded.

You are anything but that.

Steve: Thank you.

Han and I are so appreciative that you give us this time.

Steve: I am happy to be a part.

As we all are. So life brings us very special gifts sometimes, and this is one of the best. Anytime that you need to come in or have a burning question, either field it through Han or field it through me, or if you need to speak, speak. We do not want to isolate you. This is all just a process, and we are all learning as we go along.

I am going to go ahead and work with Steve, now. Han, thank you very much for coming in and joining with us today. I bid you a fond adieu, as they say.

Steve will be fine. I look forward to being with you and Steve again, as always. You should not be alarmed when I am around. I sometimes want to give you an affectionate embrace.

I don't get alarmed when I get a hug. (Therapist laughs.)

I am just happy when I am in your neighborhood.

Thank you so much. With that, I bid you a fond adieu.

Until we meet again.

Session Thirty-One
February 19, 2012

Client is inducted.

Han, you are most cordially invited. Are you available? (Long pause.)

Hello, Mary. This is Han.

Thank you so much for coming to visit with us. I hope you are enjoying Steve in his new disguise. (Therapist gave Steve a suggestion that he would temporarily have the appearance and traits of a cat.)

I am smiling. Just as soon as he is inducted, just as soon as his door is only open a crack, I open mine fully and enter.

Excellent.

I look forward to these meetings.

As we do.

Steve is curled up like a cat in the corner, asleep.

(Therapist laughs.) Hopefully, he is purring like a cat.

He is content.

Earth History

Are you familiar with Lake Vostok at the South Pole where the Russians have been drilling for about 20 years?

It is an area that is being geologically explored by the Russians, among many other scientific camps that exist there, each exploring their own area in this land of nonlocal life. *(Curious term.)*

We are concerned about the drilling because, apparently, they have finally penetrated the roughly 2.5 miles and have reached this pristine lake, which we believe has been untouched for many millions of years. Is there a danger of anything in the lake coming to the surface that might be an unknown biohazard? Conversely, are the Russians contaminating the lake in a way that might create long-term problems for the lake?

The Russian conception of environmental husbandry, or sensitivity for preservation of the natural environment, has been less than others of your culture. Yet this exploration will be of minor consequence. It should not damage, at present, the natural setting that has been preserved for millions of years under the ice there. The Russian drilling plan will be successful, with no negative consequence to its environment. It should give them a glimpse into a far earlier time, a preserved living-time-capsule of a prehistoric microbial period on your planet. The tremendous pressures of this freshwater lake will, in a hydrodynamic fashion, prevent any damage that some have feared. The lake water pressure will surge, to a small degree, back up the Russian borehole.

Are there any larger-than-microbe life forms in that lake?

There is microbial life and even larger small life existing there. It is yet to be discovered, which is millions of years old. Large diverse communities of life will be discovered there in the very oxygen-rich water environment living by chemo-synthesis. The smoking vents in your oceans, which you have discovered, will provide examples of a very rich form of life existing on chemo-synthesis, which will demonstrate to your species a broader perspective of how life exists in the Universe off-planet. Lake Vostok will be a laboratory on your planet for the study of other off-world environments, like the moon you call Europa. There are areas where life exists that you previously thought impossible.

This is a good lesson for humans to know that.

Using your present technology, exploration of Lake Vostok, will not spoil it.

Thank you. I appreciate that.

Zeta Reticuli History

How does your current leader serve your species?

I suppose we do have a leader. Our society and biology permit us a kind of confederation, by your definition. It is more a group than a single leader. They are not separate from the population. Our biology makes this impossible. Yet they focus and specialize on the concerns, leadership, and welfare of our society, as well as the interaction with other species and societies of many planets. There are the politics of understanding and knowing. This is hard to define. Often the practice of politics is a bargaining game in the human society of politics, based on what is not known. Like your game called poker, there is the "bluff." We understand the bluff from interacting with other species. In our own biological interactions, we see all the cards are on the table within our species. All parties see clearly the issues involved in their proper proportion. It makes communication more literal and quick. This often creates a form of telepathic confederation where society is fully engaged as well as the leadership. Leadership, in our society, is one of specialization of roles. I am, for instance, more a communicator and serve more in a diplomatic role. Others form leadership roles. Others are concerned with the physical welfare or medical welfare. So we do have a blending where it is difficult to discern one person's role in a society from another. Yet there are specific tasks. Even in our minds, there is a necessary need for prioritizing in role playing, each preserving the good of the whole.

The same intention.

It is just that there is a wash of knowledge that is shared by all. Your Founding Fathers of your United States knew this, as well as many other countries. Within the limits of human biology, knowledge must be shared among the population through the exchange of ideas, in news, current events, and education. We have taken that several orders of magnitude higher to a level, through biology, that would astonish these Founding Fathers of past Earth. This would be the idyllic situation in their minds, but due to the inability to be telepathic, they must use other means of sharing knowledge. Still, the intent and desire are the same. It is a good model, which we also follow, just taken to its ultimate extent. Our many discussions are about the social dynamics that telepathic communication can facilitate and the differences between us. However, this has been one of the few real differences between your species and mine. Though I wish to emphasize that we have much in common in intent. We look at it through different lenses.

You have been able to avoid so much trauma and so many difficulties.

There are misunderstandings, but less so. There can still be misunderstandings between species. Even within our species, too. We hope this has been minimized.

If I use the term 'hallmark,' I hope the meaning comes across to you. What are a couple of hallmarks of your life? For us, hallmark means a significant event or departure from the ordinary, a noteworthy point. Could you share a couple of events in your life that are especially gratifying or important to you?

I think, as a new 'born' in my species, the joining into the community minds was one such hallmark. Another was my first visit to other worlds in exploring the wonders of the Universe and becoming an explorer, crossing time and space and visiting your world. It was a new experience, which we often have available to us. The opportunity to access space and travel is much freer and more open. In a way, you are a prisoner of your own planet, but not a prisoner in your mind, imagination, and your desire to travel to the stars. That we share deeply. Soon you will be freer to explore these things, in time. Another hallmark was interacting with the species I mentioned before on another spiritual plane. In our exploration for new species, we discovered a number of species more advanced than ourselves. This helped us to see one possible future for ourselves as well. These were hallmarks of new realities, which, for each, broadens our perceptions of what reality may be.

Those are some wonderful hallmarks.

The ability to travel gives us views of physics and the dynamic workings of the Universe and a greater appreciation. It is something we enjoy, much like many humans who are well traveled having a perspective of your world that others may not have. It gives them a perspective to know what is important and what is not important.

I wish travel were a part of the early educational system for human beings.

Yes, travel is part of our education as well. We interact with other species, and we can link telepathically with many of them and see their reality through their eyes, as well as them seeing our reality. We have exercised this during visits to your Earth. This makes understanding between species better than with nontelepathic species. It requires more time and patience, such as our interactions with your species. It means taking more care when creating a nurturing contact, which will eventually fully bloom.

I know I won't be around at those times of coming into full bloom, but I can conceptualize them and appreciate knowing that they are coming.

It may not be in my time either, but it will happen.

We will be able to observe even if we are in an energy form.

On another plane, you and I will be together to enjoy this, to witness it.

We can smile together.

These are personal hallmarks that our species also share. There are few personal hallmarks because so much is shared with our species, but they are no less significant. The discoveries of the processes of a star, the dynamics of a galaxy and its workings are some of the many 'hallmarks." Some questions are for the moment, but physics is usually more concerned with the infinite. The infinite things can transcend. Even if our minds seem in the clouds, as you say, we are also very practical in what is immediate in the day-to-day. Our attention is also focused on living our day-to-day lives, in gathering food, and allowing time for peace and contentment, which is our form of meditation. These are also concerns in our day-to-day lives. Many ultimate concepts are discussed, but we also keep a grasp of our immediate surroundings.

Thank you. This is just a point of curiosity that you might be able to clarify. After our last session, I did not turn off the recorder right away, and I think you had stepped away. I recorded about 10 minutes of conversation between Steve and myself under the same circumstances, and yet it sounds like we are speaking a foreign language that I cannot understand. It sounds like it went in reverse or was Slavic or Russian. It was probably a mechanical failure, but it is a point of curiosity.

I need to explore that further. Steve is 'away' and listening to this. I am not certain about this. It could be Steve is free associating in his hypnotic state with others that I was not aware of. He may have been in conversation with others while we were having our discussion. I find that intriguing.

I didn't change any of the recording settings or distances (from the microphone) as I did the recording. It is a curiosity.

That is interesting. I hope you will preserve the recording. It may be a topic of interest in the future and may provide insights into the whole process. The communication we enjoy is facilitated in a bandwidth. There could be other wavelengths, a broader band of communication occurring of which neither of us is aware. Perhaps others of my species were also interacting. It is yet to be determined.

I will preserve the recording, and we will see what comes from it.

Does your body require physical exercise?

Yes. It is part of metabolizing our nutrients. We burn energy as an engine, much like a human being would. Our bodies are, in some ways, more efficient and simplified in their essential tasks of metabolism. We do not store fats as efficiently as you do. The need for this passed long ago. We are very efficient in burning energy. If we were to use the analogy of flight and carrying excess cargo, our bodies would need to be leaner to facilitate that.

Do you use something like isometrics to exercise or some other means?

We attempt to do that. Yes. We do physical work. We often have instrumentalities to help with this. There is the ease of technology, which can make exercise necessary in a conscious way to maintain oneself. Technology can make physical exertion easier. Exercise is physical exertion for its own sake, rather than what is necessary to perform a task. Also, our mental abilities can be so developed that our physical bodies can potentially be ignored. A proper balance of knowledge and physical exercise is important for overall health for the individual and for the culture. Our diets are low in many fats or carbohydrates, which I have mentioned. We use fruits and vegetables that make this more likely.

Can you assimilate avocados?

There is much vegetable fat in these. It is one species of plant that we can enjoy, but are less likely to for that reason.

I thought it might be a little on the rich side.

It is, but the term 'rich' is in our diet. It is one that we have sampled, and it is in our diet, but to a very small degree. It is something that would probably be, if appealing to the senses, would probably not be appealing for our overall physical health. I hope you are not offended by this. I do not wish to diminish the value you may place on such nutrition.

(Therapist laughs.) I am not at all offended. I don't have those expectations of others.

I am just the diplomat.

(Therapist laughs.) I know, and I am not easily offended.

I know this, and it is appreciated.

Spirituality

Dr. Joseph Penzien, Charlene's Father, has passed over recently.

This is so.

Do you have any sense of his presence, or is he not yet available?

Yes. I feel, at times, a consciousness that moves fleetingly. It is sometimes checking on Charlene, Steve and others. I feel that there is a consciousness that is preoccupied with many things. I feel it is learning about its new capacity to explore and travel. It does not dwell in any one space for very long. It is his nature to travel often. It does come and surround parts of Joseph's family from time-to-time. It is learning a new way, a new reality. He is in another place now that he is content with and at peace. His inquisitive mind is exploring. This is the impression I have. It is like him having a new set of clothes that he is growing accustomed to and enjoying. He remembers the past, but there is much to see in the future. He cares for those whom he loves and is loved by, but time permits only a brief visit as there are many other places to experience as well. This is the impression I have.

Thank you for your perceptions. I know they will appreciate it.

The Universe

You, of course, know about neutrinos. Are they from an alternate reality? What is their purpose?

Neutrinos are a byproduct of some physical processes in this reality. Also, they interact in other realities.

They are a part of a linking process between realities?

They can be. It is part of the quantum consciousness to create a nonlocality of communication that can permit contact across vast differences. This is part of the medium in which much of this quantum nonlocality can exist through neutrinos.

Thank you. There were some nice surprises in your answer. Does the Universe become conscious of its beauty and wonder through the consciousness of sentient beings?

The Universe has become aware of itself and its surroundings. Eventually, this consciousness transcends physics and physical time and space. Such consciousness does recognize the same values and appreciation that often unite life forms. It is like a tapestry or painting of creation that each species, with its differences, can unite in that level of beauty. Art can be appreciated from many viewpoints.

The Universe does not need sentient beings to appreciate it. What I am hearing is that it is self-appreciating.

The elements of the Universe have formed chains of molecules that become complex and begin to form simple life, which eventually grow into more complex life. This life becomes self-aware. It can then ultimately connect that self-awareness with others of similar self-awareness. It is the ultimate conclusion of the creation in the Universe. These are not the largest things in the Universe, nor the smallest, but they are the most essential. As a result of many processes within stars and galaxies, life will then create the forces that ultimately create sentient self-awareness. It is as a result of this that awareness can exist throughout the Universe.

My guess is that you have probably already visited the rings of Saturn. They are delicate. They move very fast, and they are icy. By studying the rings of Saturn, will it help us to develop a good model for understanding the formation of our own solar system?

You have discovered that there are many rings around multiple planets in your system. Other planets exist, such as Uranus and other worlds. These are part of the processes in the formation of moons. These rings are the residue of moons that have formed. One of the great beauties, which we admire in visiting your system, has been Saturn and its beautiful and unique qualities.

This phenomenon is not unique to your solar system. There are other planets with similar rings. Many of these planets tend to be large gaseous planets, not with solid cores, like the inner planets of your solar system. It is rare for solid planets to exist with rings, though some do exist with life. However, they live within a very narrow band of what we would call life. There is much life in the Universe that exists in unexpected places that you have yet to discover. Yet there is a term that you humans have called convergent evolution, which would justify much of why, in contact with other species that are associated with craft, these are often humanoid in form. This is a common evolutionary trait of space travel. It is a necessary development of technology for life to transport itself to other planets. This travel requires the ability to create tools and the need for having opposing digits to create technology to transport itself. These are examples of convergent evolution and, as a result, it is not by accident that many encounters with aliens are often with species with shapes that have a humanoid form. Still, on other planets, in the background, there exists a larger bandwidth of life that can exist in many places. Such are the planets we have discussed. Saturn is a wondrous system unto itself, and it is not unique in the Universe. There are many solar systems with similar planets, but it is not to diminish the importance and beauty of your own ringed planet Saturn.

I am surprised at the speed of the material that travels around Saturn, up to 50,000 mph.

This is a large velocity, yet these rings are very delicate, and care is taken with our craft with our gravity-based propulsion systems. Care is taken to not disturb the rings, both for ecological planetary reasons and also to not betray our presence. Our propulsion systems might leave a trace, an anomaly, which eventually might be understood.

Yes. After what I saw today on television, I can see where this might occur. (One scientist has devoted almost his entire career just to studying the rings of Saturn.) Have you ever come across a form that I might describe as an 'energy personality essence?'

Yes. We have noticed certain presences that could be an energy that has evolved from a certain physical form that has transitioned. There are others that seem to be energy in and of themselves that we do not understand, but there is a consciousness there. These could be interpreted as spiritual, and it is subject to debate.

Yes. I can see the fine line. It could go either way. This one 'energy personality' came through a woman named Jane Roberts and gave her information on reality as perceived in the dimension that this personality inhabits. It is quite an elaborate description of how reality works there and here. As a result, a number of books were written. For most people, it is too much of a culture shock to grasp what this particular personality had to say. I found it to be both shocking and useful because it shook me out of my usual belief systems. I did not mind that happening. I was curious whether you have also come across these kinds of personalities.

Yes. What Jane Roberts was encountering is not different from what we have encountered. They are often very advanced forms that can challenge the very fundamental principles of what we regard as reality. These are the questions that we ask. What really is reality, and how does it exist on many wavelengths? The existence of such consciousness is a very intriguing notion, which we explore, that touches many viewpoints or perspectives.

We question whether it is the natural inclination of some species to label it as a deity, or is it just a natural evolution of the physical world? It may just be part of the same continuum. We may be trying to look for a demarcation point when, in fact, it may simply be a continuum. That is what I am guessing at this time.

There is no clear demarcation. I would agree with this.

Have you had opportunity to visit the moon that we call Titan?

Yes. It is one of many spots.

I understand it has an atmosphere and some organic life. Is there anything else about it that is of interest?

It is a very cold place and one that illustrates the extremes in which life can exist. It is not a form of life that could form eventually into a sentient form like yours and ours. It is one where, if recognized as life, it would exist there in a limited form. Titan is a very large world, with its own atmosphere. It is very unusual with many intriguing possibilities.

Other

Steve requested a technician to come to his house to install software and a printer this last week. When the technician was at Steve's place, the technician was startled by a ball of light that whisked by him. Steve could not see it. Do you have any knowledge of that event?

Yes. It was a consciousness that was curious about the visitor in Steve's home. That was the extent of it.

He does appreciate these kinds of things.

The work was successful.

When I say 'virtual reality programs,' is this anything that your species have ever used?

In a sense. It is not technological, but telepathic. Reality is only the perception of one's environment in the mind. It can be altered for many purposes.

I know the mind has great freedom for exploring and creating realities.

Our minds, for instance, can relate very much to that. In our telepathic community, we have access to other freedoms and avenues that humans have yet to discover of other realities, other dimensions, other lives, and afterlives, which we also do not fully grasp. When our species suffers a physical death, a consciousness will continue on and transition in another form. These transitions are felt and observed by us. It is perhaps why we do not fear death as an unknown, as many humans fear death. That is perhaps an example of perceptions that we are privileged to that someday humans will also have access to. There are many viewpoints of what reality actually is. It is perhaps an extrasensory view that we enjoy that you have yet to discover, and it will fill in many blanks about reality, physically, emotionally, and spiritually.

As you know, human beings are currently exploring 'virtual reality' in such areas as 'games.' Is there any inherent danger in perfecting virtual reality programs? Perhaps it might cause an inappropriate desire to remain with these programs for too long.

It is good for one's mental health to distinguish between fantasy and reality. If one can discern, one can oblige respect and move into the other temporarily, knowing they can always return. If it becomes an obsession where they prefer to be in an altered fantasy reality, they must still exist in a physical world in a practical way. They can become detached from reality and disassociate themselves from others in ways that would not be socially appropriate. This could be a potential problem. It is important to distinguish as best one can between the two. What is reality, in the ultimate sense? We can extend this discussion forever, but in the immediate moment, knowing the difference can be very conducive in interaction with others of your species.

Open Session for Comments or Questions

Thank you. I just have one final question. Is there anything in particular that you would like to accomplish while you still remain in your physical form?

I wish to establish, in a real sense, our reality with your reality. I would like to make your species aware of ours. Yet it must be approached very carefully so that your species will not become dependent upon us due to your notions of what an advanced species would be like. It has to be done in a careful way that does not damage your species. As a result, we are injecting certain knowledge into particular individuals to raise and accelerate human evolution to a level where that eventual interfacing will be a positive one on a level where we can also coexist. This will take time. I would like to see that in my lifetime, but I feel it may take longer. Great strides have taken place recently to facilitate this, so I have great hope in that. Much of my work is toward that aim.

I am so pleased that your species is willing to work toward that aim.

Earth is one of a number of planets where this is occurring. It is an individual situation that takes its own time. It would be a time when Earth has minimized its tribal conflicts and has unified its energies and resources. That time will come, and with it will come an understanding and maturity. When that level is reached, it will happen. I have no doubt of this. It will be a full relationship, without necessitating our being discrete about who we contact. Discretion will no longer be necessary, and full contact will be accomplished.

Here and other places. I wish you well in that endeavor.

There is a Hollywood concept that, after contact, all is changed and different. This is not a true, accurate depiction. It is one of individual contacts across much time with many individuals before a consciousness is created that will, in the end, seize the day.

It is a raising of consciousness, slowly, carefully.

This is the area where there have been great strides that I commented on earlier.

I am so pleased.

You are part of this raising of consciousness. Do not diminish your part in this. You can take gratification in your role in contributing to this.

Thank you very much.

Is there anything that you would like to say to Steve before I bring him back?

We talked about human emotion and how it enthralls us and frightens us sometimes. In learning the conditions of the human species, we already know much. We are now merely working to interact with it. It is with warmth and satisfaction that I always look forward to these sessions.

Thank you very much. We also appreciate it when you let your presence be known outside of the sessions. It is kind of you.

I might pass a light across as a reminder of my presence from time-to-time. With this, I conclude our session today.

Yes. And with that, until we meet again, Namaste.

Session Thirty-Two
February 26, 2012

Client is inducted. Using a visualization technique, therapist places client on a bullet train, entering a tunnel, then arriving at his work place at 232 Main Street, 4th Floor. Client is given a large office with conference table. Therapist assigns client the task of preparing a new planet for habitation by developing plans for terraforming, including streets, sewers, water systems, energy systems, Earthworks, and preparing landing and exit sites for space craft. (All of this is done to keep the client fully occupied while the therapist carries on conversations with Han.)

Han, you are most cordially invited. Are you available? (Long pause.)

Hello, Mary. How are you today? This is Han.

I am doing very well. Thank you so much for coming to visit with us.

Steve is happy at his work.

Yes. I gave him a lot to do.

It is a lot to do. He will be occupied for a long time.

Yes.

Not as long as he thinks, however.

(Therapist laughs.)

It is a challenging task, and his knack for detail will consume a lot of time. At this point, he is somewhere else.

Yes. In an alternate reality.

Earth History

You have heard of the 'Star of the East,' the one that supposedly lead the three wise men to the birth place of the Christ child. I don't know how much of that is fact or myth. I was wondering what you might be able to tell me about what may have caused something to appear as a bright star, a moving star in the East. Was it a retrograde display?

Human attention is not continuous. It was a natural event. Yet I do not discount a more advanced spiritual element creating this situation. A star can go 'supernova,' or perhaps a spiritual being created it. We do not pretend to know everything. The human interpretation of this event created a miracle of a star to navigate these gift bearers. In the natural processes of understanding the situation, human awareness is not continuous. The situation was formed where it was interpreted this way, which created the legend, myth or reality, that you embrace. We respect this. Also, who is to say we are all knowing? We do not wish to interfere with this belief. In respect, who are we to say that we know better? I hope that answers the question. I was careful in answering this question.

I understand why you were being careful. Earth scientists created a program that enabled us to look back thousands of years to display the constellations at the time of the birth of Christ, with several years margin for error. There were some planetary alignments that may have appeared to be quite bright. As best as we could determine, it was the time of the birth event.

An alignment of planets could create a bright star of temporary fashion. The combination would attract the attention of human astronomers and the leaders who trusted the interpretations of these astronomers. This would create a temporary situation as the planets moved about in their natural order, but an alignment is temporary, and it served its purpose.

Your species has done genetic modification for a very long time. What is it that we humans do not understand about genetically modifying our food? We have been doing some of this, and it is causing health and reproduction problems.

I sense a 'Doc' question.

Well, maybe, but I think this one came from both of us.

Nevertheless, they are welcome and important questions. In the genetic modification that has interfaced with human genetics, you are often experimenting, and there are unintended effects that occur when you wish to modify foods. To create a benefit, sometimes there are unintended consequences. There is no immediate solution. It is just a continuing process of adjustment to create conditions that are beneficial to your species, and sometimes you can create foods and products that are not beneficial, but are so subtle in the long-term effects that it cannot be easily recognized at first. In time, continued experimentation creates a honing effect where it will create a beneficial situation.

Is the reaction to genetically modified foods at the hormonal level? Is that where the problem lies?

It creates a hormonal situation, which has a narrow positive effect, but at the expense of other hormones that may have negative side effects. These hormones can create an imbalance that has side effects. Where in some situations, what is perceived to be of benefit actually creates harmful effects over a broad range. I am reading Steve's thoughts. He is amused. He remembers a commercial where the warnings about a pharmaceutical drug were longer than the benefits that were listed.

That is what we call ironic.

It creates an irony where those who create this drug are merely creating something for profit and not, primarily, for improving health. It is an excuse to make money.

It may not be the intention of all, but it is certainly the intention of some.

The intention of some. He uses that as an analogy for this hormone example.

Zeta Reticuli History

Concerning your visitors, they freely come and go when they come to visit you. I know that you have communicated with them even before they come to visit. As they come into the doorway of your home, other than a telepathic greeting, do you greet them in a physical way?

We can greet one another across space and time, across a continent, or across a hallway, but we also have times of physical greeting. We do not use a human greeting, the one of raising the hand. This is an ancient, symbolic gesture showing one is not holding a weapon. We do not have that. In our form, we reach over with one hand and touch their shoulder and, for very special greetings with intimate relationships, we touch the hand and the side of the forehead, placing our palm briefly against the side of the head to show a connection or affection. This is more symbolic. In our culture, these are the gestures we use. This is our etiquette. These are greetings that focus on the mind and the soul.

It is a very engaging way.

It is an engaging form. For general greetings, we place our hands on a shoulder. For others, very special persons, we give a touch on the side of the head, very gently. They return the gesture.

Is that how you would greet those who raised you during your very young childhood?

Our early memories seem devoid of touch, but this is quickly corrected as we come into the community. Touching becomes more important for emotional development. Later, as we age, it can diminish. This is unfortunate. We consciously remind ourselves of our sense of connection and reaching out to others. It depends on the social structure of situations. It the way we conduct ourselves socially. In this social gesture, the physical connections are now muted. This is because, with our means of reproduction, we no longer extend physical touch to other areas of bodily contact. This now is done technologically, and there is a diminished emotional range. The passions and other emotions are limited compared to humans, as I have discussed with you in the past. We have observed humans and are reminded of ourselves. It is a reminder of our ancient past and what we once were. We admire this in humans. At the same time, we can be frightened by the range of emotions that are possible. Our forms of reproduction have muted in our society. However, perhaps it has also inhibited any negative emotions, to a degree. As a result, we live in peace and tolerance with each other.

It is part of the price that you paid.

Yes. There is a price. The primitive origins that each species evolves from leaves a legacy that we have suppressed or have put behind us. Yet it is always there, as much as we try to ignore it.

Is this your attempt to only emphasize the good memories?

We accept this, and it does not cause stress within us. It has been socially filtered out. Yes. the mindless primitive still exists in each of us.

When your friends come for a visit, and you plan to have a meal together, what are your customs for sitting together for a meal? What are your forms of etiquette?

In my mind, a telepathic greeting is sent out to various guests. In our minds, we work out the arrangements. There is a virtual meeting that is possible where we can send all of us invitations, and we can exchange ideas, emotions, feelings and thoughts. However, there is still the need for community and emotional connection. It is important to change our surroundings. So much comes into our minds that we could become very stagnate without traveling to visit our companions or friends. A greeting is first sent out. Then we will meet. We greet our guests at the doorway, much as humans do.

Do they bring food with them or drink?

Often food is brought. Other times it is not. There is the opportunity to express a culinary expression of the food arrangements. There is a low table in the room. We can arrange food. Some of it is prepared in a special area on a long, low, oval table, about five times as long as the width, with rounded corners. It is on a pedestal close to the floor. Arranged along this table will be various samples to try. At one end of the table will be small samples of various items. At the middle and other end of the table will be different quantities of the same foods and also something different. The gathering is very casual. Some will recline on the floor. Some will sit on furniture that conforms to our bodies. Often the chairs will seem to be in an undefined state, but, as you touch the chair, it will conform to a memory of the contours of your body. As you sit in the chair, the chair will adapt to the body. It is what we call an intelligent chair. It looks like a shapeless mass, but each body will feel comfortable in this chair because the chair changes its shape, form, and color to adapt to the tastes and contours of the figure. These are around. Sometimes we can focus our thoughts and levitate food off of this table and bring it over to us. Other times we will physically stand and walk over. The meal is often full of different flavors and textures, and there is social exchange and touching. Also, much is exchanged telepathically within this social party situation, beyond what is exchanged in human exchanges in thoughts, words and body language at a party.

Is the food room temperature or warm or cool?

Foods vary in temperature as well as in taste and texture. Some are vegetables and fruit, but there is some meat for some who wish this. It is often something that appears as meat and may not be meat. There are some domestic animals we keep to fulfill this. In our diet, the amount of meat is lower in proportion to fruit and vegetables. We have a variety of fruits and vegetables, including some from other worlds. This provides us great variety. There are some with very dazzling colors, which create visual attraction to the mind. They can be cold, at times warm, and some are at room temperature. Room temperature is relative to a human's average room temperature.

Do you have soups?

Some are in liquid. We have bowls. We have small mouths, but we have small cups, which we often raise and pour the contents into that opening in our bodies.

Do you take some time to eat a meal? More than half an hour?

It can be for several hours, but there are many other things exchanged at the same time. It is like a buffet in your human meals where there is no formal setting. There is a formal setting, but in the mind. In the room, no one is physically sitting at the table together. In social interaction, you would expect closer contact at a table in a formal setting. In our gatherings, this would appear to be not so to a human. There is that mental connection that brings the same thing. It just appears more disperse when comparing it to human gatherings. Our guests are in interesting conversations and exchanges. We often share the sensations and pleasures that the others are experiencing. They can sense the taste and flavors of other foods. This is exchanged. We can taste the other person's food through the thoughts of the other. It combines into a feast of the mind that may even extend, if there are limited food supplies, the enjoyment that extends way beyond the amount of food that might be available. Each mind has its own interpretation. This exchange can be interesting and humorous, as each participant shares in the enjoyment of the food, and each has their own interpretation.

It is interesting how humans all have taste buds, but their reactions to taste can be quite different.

This is true. One can experience sweetness or sourness. They know what sensations are being experienced, and yet each individual mind is interpreting the taste with their individual taste. In my society, the expression of different tastes and reactions to various foods makes it easy for the host because the host knows what their guests enjoy or do not enjoy. Their favorite flavors are always present and available in the mind. It is another advantage of our telepathic abilities.

It helps me to understand this level of detail, and visualize them, because I hope that I will be able to visit in spirit, at some time.

The room shows several levels of floor surface. There are curved walls and ceilings that blend together, forming majestic arches with no corners. It is expansive. The ceiling is not low. There is a sense of expansiveness. I see greens, oranges, greys and some brown tones in pleasing hues. Some of the walls and different spaces can be changed by thought, depending on the mood of the host or the guest. They can change a color by thought, expressing the mood of that particular individual on that day.

I find that quite interesting. Colors in a room affect me.

They affect us very much. It is why a house can appear different each time. It is the same physical structure, but the colors, the light, the mood, and the impressions can change.

What are a couple of the colors that Zeta tend to agree on for a comfortable space?

The Earth tones, as you call them, are particularly calming. Browns, greens in various light, beige and whites all create a soft light in the eye. This creates a gentle and appealing environment to their senses. There are other, different spaces in this dwelling. There are also many windows or what would appear to be windows. These are openings that you can walk through, yet they can create a field, which is like an invisible window that you can pass through. The environment can be held out, if there is inclement weather, which we have from time-to-time. The external temperatures and internal temperatures are different. There is a sense of openness and yet a feeling of enclosure and security from the elements. We have a sense of space and belonging, which we create in our minds.

The light that comes in through what you refer to as windows, do you create an artificial sunlight to come in, since you are underground?

There are natural sources, and there are artificial sources. Sometimes these are in combination to create the mood that we wish to have. Sometimes these moods are established by the host. Other times they are established by the guest. At other times, they are established in combination, creating what we call a 'crowd' mood. It reflects the collective thoughts of all participants. It can also indicate the mood of the party. Some parties (or gatherings) can be made very successful by the colors of the room. Collectively, the intelligence of the room will interpret, with our psycho-technic connection, and do its best to adapt to the collective mood of the whole party.

So the room is reading the mood?

The room is reading the participants in the room. Often there are presets. The host can help influence the thoughts by selections of certain colors. If some of the guests are in a dark mood, the programming of the room can be prearranged to ease those thoughts to make it a more comfortable and appealing environment. The room can be proactive in creating a positive experience so that a black mood does not permeate the space and affect the other guests.

Yes. I can see how a black mood could easily permeate the space. (Therapist was surprised by this reference to dark or black moods.)

Yet there is a healthy mental perspective that, in the programming of these structures, we anticipated and have made appropriate corrections for potential dark moods. These are details that are built into our structures for interacting with our society. It is not to create a kind of artificial reality, but to reflect reality and put it in a positive light.

Have you experimented with something akin to stained glass to reflect various colors?

The glass tends to be virtually nonexistence. It is more like a force field, but the house or dwelling can create windows that can change in color and in visual tones to appear as stained glass, for example.

It can throw the colors into the room in shafts of light?

Yes. There are also skylights. These dwellings are often sliced into the hillside. They blend in with a large swathe of windows. The tops of the windows are hills of green grass. There are also openings for natural skylights and ventilation. These can be very appealing. Natural light is important to us for our physical health, as well as our mental health.

Do you also have a strong need to be in sunlight?

Yes. This is for our emotional health, as well as physical health, and our need for elements, vitamins from the sunlight.

Vitamin D?

Yes. This is important to us, too. The artificial light does not provide these. It comes from our stars, which we orbit, and they provide a bath of light.

We have recently discovered that the Vitamin D from sunlight is critical to activating many enzyme-based reactions in our bodies. Our use of sunscreen has actually created a shortage of this vitamin.

We have adapted out bodies to the atmosphere. Most harmful rays are filtered out, much like your ozone layer. You do have an ozone layer. This not uncommon. The ozone and similar elements, which provide filters, fracture the rays and disperse most of them harmlessly. We also have ozone. This is a natural part of many planets, particularly ones with life, at least the humanoid type of life to which we are accustomed. We both share the effects of sunlight in activating enzymes and mineral reactions that help us to have healthy, long lives.

When you look back over your long history, could you name two or three beings of note who have contributed to your species in ways that have enabled your species to survive and to thrive? They could be philanthropists or humanitarians (excuse the term). They could be scientists. They could come from any field.

Our scientists have made significant contributions to space travel. They have opened up the Universe for us, not only in a physical sense, but also in interdimensional travel. We have left monuments to the individuals who enabled these breakthroughs. There are also social pioneers who have challenged the ways of our telepathic minds in requiring full consensus. Instead, we have a 'grey' field of everyone, with only partial consensus. We introduce new ideas to stimulate our continued development. These will often challenge our sense of social patterns. There is a desire that each lives a productive and contributing life of value. These are beings who have contributed both individually and for the collective good. You are reminding us of that relationship, that we have an obligation to each other. *(Zetas do not have a hive mentality.)*

What word could I use instead of humanitarian that would convey the same general tone for your species? It would be that group of individuals who desire to promote the welfare of the whole.

These beings are concerned with the obligation of welfare of the whole so that all are content, and none are deprived. With the mental abilities to understand each other, it is hard to hide one who is suffering, which can occur due to heath reasons, or the effects of living in remote areas of our planets. Each adapts to the environment of the immediate surroundings. We try to create a universal society, but these do not happen by accident. It is necessary to create conditions, which are appealing to each.

One individual in particular was courageous in reforming our society by implementing a strict control of our population. He persuaded our population to see the benefits of this. This was to create a balance between the collective whole, and yet remind us of the individual and the balance in between. As a result, we live on worlds where there is no stress on our planets, and there is plenty for all. These have been great social patterns, which did not come easily. In time, we remember some who stood out in implementing these goals, which have been achieved. As a result, we enjoy the benefits of realizing these goals. We continue these patterns on other worlds to provide comfortable living conditions, thus creating good homes for all the inhabitants on many worlds. We often terraforming planets to achieve this. Also, there is the terraforming of social patterns. Both are essential for a successful planet.

I failed to give Steve the instructions to consider the social patterns in his task assignment, but that will come.

It will come unconsciously, as it will naturally create a planet that is appealing for all of its inhabitants. This will evolve. This is one of the things that will be learned, the social terraforming.

A very good term.

It requires discipline, tolerance, and compassion, as well as knowledge and awareness of the conditions of others. The telepathic abilities make this much easier to implement since each is aware of the other.

It cannot be underestimated what telepathic abilities we have. They create much insight that one day, hopefully, humans will also enjoy. It provides many perspectives about the society and about life in general. The early impressions of new life are shared by all. The insights that are provided at the end of life of an individual create a belief system in an afterlife that one does not fear. Our concepts of the soul emanate from this telepathic ability, too. There are more impressions that are shared. The structure that results from this cannot be underestimated.

Thank you for that.

Our party will often conclude with guests departing and arrangements made for future parties. The guests and hosts of such parties will continue the gathering into the evening. We have a sense of light and darkness. Like on Earth, we have evenings at the end of the day, which tend to be the traditional meal times. Friends will part. We often exchange touching. The departing gestures are the same as the greeting gestures. This is much like the word 'aloha' in your culture, which means the same for arriving and departing. This is also true of our physical gestures. They have the same meaning, coming or going. Often a party will conclude this way.

Thank you. I feel like I have visited one of your parties due to your descriptions.

I wanted to share how such social gatherings come to an end.

That was quite appropriate and an excellent way of summing it up.

We would seem to not be as physical at times. At other times, we are close. We would stand in a large room, yet there is a close connection, due to the minds that are present. An intimate party could be in a large room with few individuals. Yet we also see the need to sometimes be close.

I think the touching, at least for us, also changes the brain chemistry. It produces, as they say, 'chemicals of happiness.'

These are not foreign to us. They do exist in us, also.

Spirituality

This morning, I went to a church service, and I invited you to step in and experience a room full of people who were surrounded with a number of spirits. I don't know if you had the opportunity to step in for a moment.

There is a symphony of thought in these situations. I was present. Thank you for the invitation. There are elements of a spiritual sense that are not often fully apparent to all who are present. Everyone in the space is still enjoying the spiritual interaction in various degrees. There is a radiance and interaction of spirit and physical presence. There are more energies swirling in a room than most would realize. This we experienced with those who were invited. We can see the spiritual energies moving about the room. It creates an ocean of positive energy and of emotional love, which creates a sense of well-being. This is necessary from time-to-time. This creates a support system, which humans have determined to be important. These are healthy actions that benefit all who partake in them.

Thank you for your comments. I rarely see them, but I feel spirits coming around.

Sometimes the presence of a spirit will pass like a breeze. You do not see where it comes from or goes, but you can feel its presence passing through you. I, in thought, sometimes may appear in this fashion. Sometimes you can feel the gentle pressure of a touch. We have talked in the past about this. When it is recognized and understood, it is one that is enjoyed.

Yes. I do enjoy it.

We enjoy it, also. This interaction can send warm greetings and support, and it is important for us. It is important in the physical elements, as well as the spiritual elements.

It has also enabled me to understand that, if I send a thought to you, and you have the opportunity to respond, you are able or willing to do that.

I am often willing. It is not in our composition to just ignore requests. I am struggling for the word to describe what requests we are to ignore, though we can filter or block out, because otherwise we would be flooded with many thoughts. We can focus on a single voice in a large room and, at the same time, we can see many other elements in play. We cannot just ignore anyone we wish. We understand that others are reaching out, and it is part of our etiquette to respond in kind.

I enjoy that aspect of your culture, the courtesy that you exhibit. I appreciate it.

Many who have experienced abductions by other species might find our behavior incompatible with our other actions and behaviors. We anticipate this. Understand that, as in humans, there are many personalities and behaviors. We often create a culture in which we wish to create a connection and develop a positive relationship. For those who have experienced negative experiences, please understand that there are other species where the interactions are very positive, or they wish them to be.

George (a mutual friend) could certainly comment on that.

Yes. It is a mixed group. Know that we protect those who are vulnerable from those whom we know would wish to exploit or abuse them. Those who abuse are more the anomaly than the majority.

That is good to hear.

There are, as you say, all kinds.

I had no way of analyzing if alien visits were mostly well-intended. Thank you for sharing that perception.

Well, we wish to think so.

Contact

May I ask you about the Leed (pseudonym) family?

Yes.

There is a Thomas Leed, a son named Chance, and a brother named Matt (pseudonyms).

Yes. Matt or Matthew.

Apparently, this family has had a long history of interaction. I am specifically asking about a number of mysterious deaths associated with this family. A Dr. Michael Buckner, the Father, a partner of Thomas, and a psychiatrist, all died suddenly. I think most, if not all of them, died from some kind of viral condition or unknown condition. There was a military presence involved. I am wondering if these deaths really were natural.

They were designed to appear natural. The causes of these deaths could be natural, or they could be influences from the entities that are visiting them to suppress information and dispersal of general knowledge concerning the abductions, which these energies are conducting. They wish to conceal themselves. They do not wish the abduction phenomenon to be well known in human society. It is best seen as something that possibly cannot happen. Such an attitude held by humans is helpful to them.

Well, Tom or Thomas is planning to talk on the radio, so their efforts to keep this information secret may not be working for the entities.

There was a comfort level with the entities interacting with them. They did not want the exposure of information. Plus, there was some harm done. The psychologist was well meaning and a good person who was effective in treatment. However, his full effect was not permitted. It was denied and, in time, he would have offered more beneficial assistance, had he been permitted to continue his life. His life was stopped short by these negative energies. They killed him because he knew too much.

Was there interest in this family because Tom and Chance have that unusual ability to affect electrical things around them? That is one thing.

They are 'super' human who are developing early telepathic abilities. It is a continuing process of creating a prototype of a blended species that can exist on your planet and other planets.

Where did the prototype begin with that particular family?

Many generations back in the great grandmother and the great grandfather, grandparents, and then parents, then his sons and brothers. The young son and twin daughters also show the fruit of this continued process. Something has been introduced genetically with the son. There are many others on your planet who are also enduring similar kinds of programming.

There is much more going on than I had imagined.

The advanced abilities the humans have been given help the entities to interface with them more readily. It makes the process easier. Also, when contacts are made, it makes the contacts more comfortable and efficient. The exchange of thoughts is more readily possible by downloading information and impressions. It creates insights for these entities in understanding Earth social patterns and experiences. They are very interested in the experiences that humans have so they can gain understanding of the human species.

Hopefully, these are not the species that we discussed last time.

The Verdants. The Verdants are present. Their agendas can seem selfish. They wish to overwhelm other species. George has commented on this. It is not necessarily a good thing. They are not invasive, but tend to overwhelm others across space and across time. They are like a species of spores. They envelop and change. They are not necessarily a threat, but they do wish to dominate. They have an arrogance that makes them feel superior. It is in their attitude, and they are not as superior as they think.

The attitude of superiority is a big loss for them because that prevents them from perceiving and learning what might be a much greater view of reality over the long term.

They do not care for balance. They wish to unbalance nature for their own benefit to create situations beneficial only to them, not integrating into the natural environment. They use the old natural forms of sexual reproduction to intentionally overpopulate, which stresses planetary systems. This forces expansion for continued resources. It is not a system that is one in balance, but it is one that promotes expansion, which is not necessarily a good thing. It is an ominous thought. We have attempted, with other species, to contain them. They are not as advanced in the evolutionary sense. This may seem arrogant, but they are not of a moral quality that is appealing. We look down upon this, yet it may seem we are the same for thinking that way. However, there are certain obvious things apparent to all in the Universe that define right and wrong.

There are certain universal rights of life that must be respected and not selfishly exploited. The Verdants are ultimately such a bad species. They invade, appearing to be a blessing, but this is ultimately a disguise. It is much like an occupying force where change is made to accommodate the occupying force, but they only create a dependent population and use other species for their own aims. We are talking now about the more negative things, but this needs to be discussed, also. I have often painted a bright picture of the Universe, but there are elements that are dark, and their purpose is dark.

I am aware that you have experienced darker spiritual energies, which you are successful and adept at repelling. I believe you understand some of these elements. You are aware of the Verdants, as well.

Awareness is the first defense.

Yes. We are aware of them, also. One reason we are connected in this way with humanity is because we share a similar desire for a positive outcome. You are unaware of us, but we are aware of you.

You are unaware that we are allies in the struggle for what is positive and in harmony with the natural processes of the Universe. The Verdants are not natural. They can be ugly in their lack of morality. I am trying to be cold and objective. I am just stating the facts. I could let passion enter. This is not the way to combat the disruptive nature of this species.

I don't think we should give sway to our emotions when it comes to thinking about negative energies.

Exactly. You understand. I am not going to be swept away with passions, but know that we are well aware of the activities of the Verdants. We provide a strong, broad front to slow or push back the ambitions of other species, which we feel inflict damage to the universal sense of life.

That being a universal truth that life should thrive as long as it does not push out other species?

Exactly.

The Universe

Why is Triton moving in opposite orbit of the expected orbit for Neptune?

Neptune was struck by an object, which was absorbed and created the orbital rotation path that is not consistent with the other planets. Triton orbits in a polar alignment, which is not on the plane with the other orbiting planets. The planet originally rotated in a pattern where it reversed itself, yet the moon (Triton) continues to orbit in its original path. The planet suffered a cataclysmic collision, which changed its alignment. The planet survived as well, but the effects of this collision are still evident. These are things yet to be known by your species because you would not have anything to compare it with. In time, as you explore other solar systems, you will see new things that will enlighten you about the origins of your own solar system. You do not yet have the experience of interstellar travel to put the many things you observe into their proper perspective.

I suspect that we are not looking at a large enough view of the interactions.

This is only natural. In your scientific process, you can only know what exists based on what you observe. Your powers of observation are confined at the moment, but are expanding rapidly. It is only a fair assessment. It is part of your scientific process. You can speculate, but it is based on facts, which you observe. It is only to be expected that you would form certain conclusions. As you log what you observe in the Universe, you will see that this is an example of what would result from such a collision.

Of those things that affect the Earth, its orbit, its temperature, etc., to what extend do these effects expand beyond our own galaxy?

There are conditions of space that exist in your solar system and beyond your solar system. You understand this, and you are sending out probes at what is high velocity for you, but it is painfully slow for us to observe. In time, you will learn the border between the influence of interplanetary space with the environment of deep space, and you will see the contrast that exists. You will see, in perspective, what influences your solar system. The void of deep space is quite empty. The spaces in between the local solar system, by comparison, are relatively busy, with many elements interacting. There are many comets, asteroids, and elements that are part of the solar system. There are islands of activity in between the vast emptiness of space (which is not truly empty). There is dark energy and dark matter, as well, which exists between these solar systems. You are beginning to discover this.

Thank you so much for that. We have recently discovered a water planet.

Yes. It was spotted by your Hubble.

I didn't get anything on it other than the discovery. I don't know if they meant entirely covered by water or mostly water.

It is mostly liquid water.

For us, the unusual aspect is that the water is in a liquid state.

It is in the 'Goldilocks' zone, and it is a larger planet of oceans. It is one more example of the abundance of life that you are slowly discovering. The recent advantages in technology are creating a golden age of discovery in astronomy that we take for granted. It is new to you. We delight in your discoveries.

Are most of these oceans on other planets salt water?

Mostly. They have various degrees of alkali. Water is more abundant than you realize. It exists in various forms in the Universe. Hydrogen and oxygen are common elements throughout the Universe. Also carbon. There is a combination of carbon-based life, from which we ultimately evolved. Our silicon modifications are only recent in our evolution. Most life that forms is carbon-based and combined with oxygen and hydrogen to form water. This creates a beneficial nursery for life virtually everywhere, where the conditions are correct. It is almost inevitable. It is almost hard for life not to exist.

Regarding this water planet. I am going to assume this newly discovered planet has sea life. Is there any kind of land life?

The life created early in the oceans eventually, like on Earth, moved up onto land surfaces, what land surfaces there were. Many land masses exist to varying degrees on these water planets. Our planet has the same land area as your planet, but our planet is larger, almost twice as large.

I think that is a plus.

As a result, our oceans are larger, but this provides us with the oxygen-creating elements for our atmosphere in great abundance, so we have a thicker blanket of abilities to support life as we know it.

And good air to breathe.

Good air to breathe in the form of respirations. Our population is strictly controlled, and we are interconnected with our planet in a spiritual and physical sense. Much like your Native Americans respect the land, so we also respect the land on our planet. We have many of the same values, and we have a planet that affords us a great luxury of abundance, a kind of garden.

You probably know that I love gardens.

Yes. I choose that word to help in visualization. Often in our communication there is verbal exchange with words rather than complex images.

I can well imagine that it would be confining for you.

Yes, but do not be concerned. It is just part of the interaction that we enjoy. Sometimes impressions can be placed in you in nonverbal ways. I think you are aware of this.

Yes.

Other

I realize that you are reading my mind.

At times, I can anticipate your question, but I repress this because it is not good etiquette.

It also helps me be part of the flow for us to stay in this mode.

This true. It is patient to wait for a thought to be fully expressed in a way that is appropriate for your species. It is necessary (to do this) to create a rhythm of communication.

Yes. I knew that you were waiting, and you knew what was coming, but it helped me to get it out.

Understand that you can sometimes surprise us. We delight in the unexpected. It creates an interesting relationship.

Thank you for that.

Open Session for Comments or Questions

Is there anything you wish to add before we close out our session?

Steve is content. He is preoccupied. You have given him much homework, which he is enjoying. I am experiencing the human condition when I interface the thoughts between you and your species. It seems now a strange word to use (species) as a term for you.

I also have difficulty with that word now.

It is an inappropriate word now. This is a result of new understanding. Do not be offended.

I am not. (Therapist laughs.) This is a transitioning point for both of us.

You are not offended. I can sense it. There is a new level of maturity, which we have commented on before. A greater understanding, and the word 'intimacy' is a sexual one in context, but this is not, indeed, the case. It is the only human analogy that I have. It is a maturity that comes from, in your species, a relationship between adults, when it is in full measure. Yet this is not an appropriate one for our interaction today, yet it is the closest in comparison. It is a relationship of intellectual thought and also of an emotional and spiritual connection that now exists. I hope I have expressed this well. This is my observation. The term 'species' is now inappropriate.

It is too distancing.

It distances. Perhaps a new word will have to be created. We can create one that will reflect the proper meaning. We will have to invent one.

Yes. I appreciate the invention of new words.

This is an observation and a question. In this new experience, we need to find that word which characterizes that relationship that now exists. It is one of maturity and depth, which continues to become a significant and important one. It helps us in understanding your species, as you are beginning to understand ours. It is one question that I will continue to search for the human word that expresses what we are both seeking to understand.

I am fine with scanning other human languages to find a word.

It is a question to you, but it is also a question to me so that, together, we can both find the word that best expresses this new understanding.

And connection.

Thank you. That is a good word. A 'deeper connection' might be a phrase to describe this. We will move through this and design a new word in the near future for that question, if that is appropriate. I am available for further communication, if you have questions.

I am sure I will have more questions.

Steve is tired. There are limits, but we are communicating on multiple levels, which is natural for us, but can be a bit of a strain for humans, particularly in the interaction.

Please come in our dreams, if you can.

With Steve as our translator, the 'hose' that the 'liquid' passes through is only of a certain diameter. Frequencies of communication are expended beyond normal human connection, but it is still one that is narrow for us. Still, it is what is accepted to facilitate this communication today. With the strain, which is only to be expected, he is performing his task well.

With that understanding, we will see you in our dreams.

We will, and you will feel my spiritual energy so that you know that I am in the neighborhood. It pleases me that you are not alarmed by this.

Not at all. (Therapist laughs.)

It is just a lotus blossom fragrance that passes and will come again when it is appropriate.

Beautifully said. Thank you and adieu.

(This therapist gave some thought to Han's reaction to the word 'species' when we referenced each other's species. He wants a word that will be less distancing. After some thought, this therapist came up with the concept of 'Zeta kin' to denote a closer relationship when referring to Han's species. This is to be discussed at a later session. It is a kind of offshoot of the concept of kindred spirits.)

Chapter Two

Session Thirty-Three
March 4, 2012

Client was inducted. For the second time, using imagery, this therapist placed client on a bullet train, entering a tunnel, going to work at 232 Main Street, 4th Floor, in a mythical city. Client was given a large office and conference table. This therapist assigned client the task of preparing a new planet for habitation by developing plans for terraforming streets, sewers, water systems and preparing landing and exit sites for interplanetary craft. Client was given two assistants: Scala and Lent. Client provided the name of receptionist as Ms. Peters. (This was a hypnotherapeutic technique to keep the client's subconscious mind engaged in something else, while this therapist communicated with Han.)

Han, you are most cordially invited. Are you available? (Long pause.)

Hello, Mary. How are you today? This is Han.

I am doing very well. Thank you so much for coming to visit with us.

How are you Han?

I am doing very well, also. Steve is happy at his work. *(This is a reference to the therapist sending Steve to work on his project of terraforming a planet.)*

Earth History

You know about the Neanderthal man and the Cro-Magnon man. Were DNA samples ever taken of them and placed on other planets?

No. They were earlier stages of man, prototypes you might say. They were not transplanted. They were merely steps toward a later version of humans. These later genetic versions were transported to other places. The Neanderthal was left on its original home planet. It was just an evolutionary step to a later being that was transported.

Could you tell me what Earth species was able to form words and name objects and when this happened?

This happened soon after the Neanderthal, some almost 500 thousand years ago. This being began to create words and verbal communication. This was followed by more articulate sounds. From areas inside of what you know of as Africa, these species migrated around the globe. Many land bridges existed then that do not exist now. Dispersal of these species occurred far and wide. They existed in a number of climates, from the equator to your Polar Regions. They could tolerate extremes in heat and cold. They predominantly lived in the mid regions in what is now Asia, Europe, and North and South America.

What about Africa?

Africa was the point of origin, and they maintained this as their parent continent, as well.

When we look back at the history of the Earth, I believe the Earth was impacted, and it threw up the debris that eventually made our moon. What part of the Earth would indicate where that impact took place?

Resulting in your moon?

Yes.

The impact occurred when not much of the Earth's crust was in solid form. Most of the Earth's crust was still molten. There was an impact point, but at that point, the surface was too fluid to be any relevant reference point in current times. It was like two drops of water, one parting from the other and forming a mass of the same material, which eventually created a stable orbit. It was in a time when both bodies were in a gestation stage of solidification. In time, the orbit was stabilized, as you know it today. These objects began to evolve differently. Your moon would never support an atmosphere, due to its mass, but it became a relatively static object, while the Earth continued to evolve and create an atmosphere, due to its larger mass. It was able to create the planet that you know today. I am relaying something to you in a phrase, which is almost an inconceivable length of geologic time.

Is the planet Earth older than we think it is?

Earth scientists are getting a good grasp of geologic time approaching accuracy. It is on the order of billions of years. The moon is a long and faithful companion to its larger partner world. As you know, it orbits in a synchronous rotation. The characteristic orbit is a result of the early impact when the Earth and the moon were still in their early molten stages.

You remember Leslie and Spencer McDaniel. I got a rather excited call from Leslie regarding internet information about there being deep, pulsing sounds being heard in various places around the Earth. It has been going on in Norway, the United States, Hungary, and Canada.

Mystery sounds?

Yes. People are experiencing deep pulsing sounds that seem to come from the sky and cause vibrations around the Earth. It has been going on at least since last March and includes Kansas, Russia, Canada, California, etc. The sounds are deep, sonorous, and pulsing so loudly they cause the buildings to vibrate. A scientist says these are a result of corona mass ejections or interruptions in the radiation belt around the Earth. If this is so, why isn't it just being heard at the poles? Why is the sound dropping down lower (from the pole) into Europe? If I were to guess, it sounds like some of the sounds are metallic and perhaps like the effects of an extremely large craft, but I don't know what is going on.

Some of the sounds heard relate to the coronal effects of the sun, but this would be an incomplete answer. It is predominately the reverberations of these, which are fairly disruptive to human ears. We cannot detect these sounds with our own ears because our hearing is less than humans can hear. Part of what is heard is the vibration of the planet that results from the coronal activity. This has gone on since the planet was created, Now, certain areas of the planet have become more sensitive to these sounds, particularly to the populations. It is a natural resonance of the planet and is due to the electromagnetic effects of the sun and the magnetic poles. These energies will radiate out from the North and South Poles and permeate through the planet. Sometimes they are quite random. They will emanate in certain areas. Many of these also radiate into areas where there are ocean regions. The propagation of these low-frequency sounds through the water is sometimes magnified by the oceans. Magnification through the water is more efficient and faster than it is through the atmosphere, and the large bodies of water will act as amplifiers to magnify these sounds at certain times and places. It is more or less a new phenomenon to humans, but it has a natural origin. People have become more aware of its presence due to the introduction of new technologies.

I don't know if this is only for ultra-low frequencies (ULFs) or all sound transmission. Why does it pulse rather than just emit long tones?

It is due to atmospheric cells of varying density. These are the varying densities of turbulent air caused also by the energy radiated from the sun. It is correlated with how these vibrations resonate off of the planet in an extremely-low-frequency sound/pulse. Sometimes it manifests itself this way.

We call them ultra-low frequencies.

Extremely ultra-low frequencies. The sun creates a large volume of energy that is quite massive. As that energy touches the planet, it causes the planet to reverberate very softly in an extremely low-frequency, occasionally noticed by humans.

Yes. That makes sense to me.

Part of it has to do with the characteristics of each planet. Many other planets can exhibit this same effect. It depends on the composition of the planet and the extent that the molten core allows vibrations to occur. This is dependent on the molten magma and movement of plates in a complex interaction. These vibrations will resonate in different ways. Often these sounds will appear and disappear rather suddenly and randomly. It is much like the sun playing a flute (the Earth) and the 'flute' is shifting and covering certain holes to change the frequencies to create different sounds. The molten magmas are the fingers playing the flute and creating these random effects, to use an analogy.

It is a lovely analogy. I think it also explains experiences of people in the past who described hearing the angels playing trumpets. Some of the sounds that I heard on the recordings did sound like very large, low-sounding trumpets.

It is a remarkable trait of humans that, although they may lack a scientific explanation, deep within your minds, you can intuitively grasp the very things that we are discussing. The songs of angels from the sky are not far off in human interpretation. The answer is much more sophisticated. Never-the-less, they are not far from the answer already.

It is so exciting and fun for me, but it is frightening for some others.

Some humans are more sensitive to the sounds, and they can become distressed. In an anxiety-laden, modern society, becoming aware of 'new' events can magnify that anxiety. It is the Earth singing, resonating from the coronal effects of the sun.

That is a good way to describe it.

Though many are not accustomed to these sounds, they can become anxious. It is a natural phenomenon. It is not an ominous sound of the end of the world, or of any technological development from man. However, the propagation of your species, in covering the surface of the planet with more structures, also changes the tone of the 'flute.' Modern cities have, you might say, changed the 'reed' in the flute. The mass of the cities gives it a different tone. All of these are factors.

Is this part of what we describe as 'music of the spheres?'

You can say this as I understand your meaning of it. It is the sun radiating out energies that vary in many wavelengths. The Earth is considered a dense and heavy planet. The electromagnetic fields that combine with the Earth's strong magnetic presence, and the profound magnetic fields create these resonances within the Earth. Yet it is a minor inconvenience when one considers the blessings that this magnetic field has created in protecting its surface inhabitants from the harmful radiation from your sun. Without this field, no life, as you know it, would exist on your planet.

I will use the analogy of the flute to explain this to Leslie.

I smile. I can imagine the image of my appearance with me playing a flute.

(Therapist laughs.)

It would evoke many old mythologies, Greek and Roman mythologies (Pan) which are amusing and romantic.

And the American Indian had them, too.

Thank you for reminding me. They hold a special place with us. We are very fond of their cultures. Our cultures and their cultures are not dissimilar in regard to how we interact with our respective planets. They treat the planet as their source for life, like a parent, the Mother of them. We regard our world in a similar fashion, and it is with affection that we regard their culture with great respect for the insights and great truths that they hold.

Yes. I feel a lot of kinship with them.

I am surprised that more humans don't let their curiosity take charge outside of their belief systems. If they do, I do not know about it.

If they do, within human cultures, and it is not acceptable due to a certain order or dogma, it can be considered heretical. One could risk becoming an outcast.

Even now, strangely enough.

Our telepathic abilities greatly diminish such thoughts because they are shared by everyone. There is a mutual understanding, and no one is an outcast at that point. We all see, understand, evaluate and accept or reject certain concepts. The general summing up will guide us in a scientific fashion concerning new discoveries. We have the strength of the many in intellectual thought in exploring these things (new ideas) within our Zeta community. Those who risk different thoughts, because individual thoughts cannot be shared at a level of understanding in humans, can risk becoming outcasts. Many outcasts are not lonely very long. Often they form new beliefs. As a result, new pathways or directions are taken in many cultures. The new beliefs hold sway until those beliefs systems are challenged, once again, by new outcasts. On and on it goes.

Yes. Deviation is the growth factor.

Yes. Your Darwin knew this, and there are similar patterns in cultural beliefs, too.

Yes. I have seen social change just in my lifetime.

There have been many new changes in your planet. I look at the beliefs just 50 to 60 years ago. There is a new sense of tolerance. There is a new level of understanding among more humans. Many bad traits that cultures once accepted are now being abolished or pushed back. For example, there was much more slavery and the subjugation of the females of your species. The greater understanding and tolerance within cultures is most promising, including a higher level of communication. Also, the struggles in your world have brought together many different cultures, sometimes in a common struggle. There are the struggles of world wars or the desire to relieve famine and hunger. There are other struggles for intellectual development and education. These all combine in a good and positive way to raise your level of civilization. You have much left to do, but you are making great strides, and we find this encouraging.

Thank you. I have heard it said that physical life is, by its nature, ultimately doomed to extinction. Scientists put forth the idea that most of life that has existed on this planet is now extinct. There is a difficulty in all of this. If we are evolving in that manner, we are not finding what I would call new interim life forms. If there is a long path of evolution, I would think that Nature would continually put out new forms of life to see if they would survive. Would you help me understand this process?

Evolution, from the perspective of each age, appears to be slow, yet life finds a way to survive. Sometimes cataclysmic events can intrude, wiping the slate clean, such as your dinosaurs, which had a very long and successful biosystem. It served sufficiently for the time it endured. It also provided a basis for what was to follow. That life was destroyed, but the basic elements that created that life continued to exist. These basic biological elements, at the level of microbes, can recreate and form like clay into new shapes through genetics. The one thing that we have noticed is that life is irrepressible throughout the Universe. You wonder if you are alone. You are most certainly not. You are like an aquarium in your own isolated glass box, immersed in an ocean of life. Due to your primitive technology, you cannot yet pass through that glass of the aquarium into the larger ocean.

You would see a great expanse with much biodiversity. Life is irrepressible and, one day soon, you will understand this. You learn in small steps. It is much like our contact here, for example. I admit that I give you hints, sometimes with intentionally vague answers. This is only to not harm and stymie your own development by giving you too much of the answer to the equation too quickly. Also, this allows you to raise yourself so that when we do meet in a formal fashion, you, along with many other forms of life and societies, will meet us as independents and will not be dependent upon us. This will take time.

Maybe that is the problem. Maybe it is evolving too slowly for me to observe in my own lifetime. I would have expected to see a few mammals coming out of the ocean by now, but my lifetime is short.

It is evolving faster, but still at a pace, both within your lifetime and mine, that still seems too slow to perceive. Its rate is almost imperceptible. Our minds have become so full of knowledge that we have anticipation that cannot be fulfilled fast enough. Although we learn faster, we are not capable of watching the evolution pass before us. Each event moves at a much different cadence.

That would explain it to me.

It is moving faster, as you have observed, but still too slowly for our intellectual observations to grasp. This has more advantages than disadvantages, both for the introduction of new species or the artificial genetic development through technology of such species.

I will just have to occasionally stop by in my spirit form and check out the progress.

Your spirit form will accommodate that nicely. You will enjoy it.

You know the Aborigines in Australia. They have a 'dream time' and also what they call a 'walkabout.' Can you tell us about who might have interacted with them in earlier times? They have unusual knowledge about astronomy. Perhaps that is why they have a 'dream time.'

We spoke in earlier sessions of hallmarks of lives. One hallmark for the Aborigines might be their 'walkabouts.'

Yes.

There is a new realization and understanding, as they move away from their offspring and go on a journey on their own to find out about the world and themselves. They gain an understanding about the world, and their place in it. Encompassing this are the stars and the sky and their intuitive sense of the great vastness of the sky that goes on forever. This is a sky that all life shares, not only on your world, but throughout the Universe. No matter what life exists on any planet, we all share one profound connection. We all share and wonder at the sky, the stars. The Aborigines are capable of higher sensory perceptions that can accommodate life forms such as us. They are more accepting of entities, such as us, and we can visit them. Our visits can be more overt. There is physical interaction with two other Earth societies where there can be a natural and free exchange of thoughts. As a result, these cultures see the sky with a different perspective. Uncontacted societies may see the Aboriginal view of the sky as non-scientific and less sophisticated, but this would be in error.

They have a spiritual sense of the sky, as well as a physical one, both of which are sharp, clear and astute. They see the energies that pass and incorporate these into their culture. They regard the land as their Mother. They may consider the sky as their Father. Both are inseparable to their concept of how they perceive themselves in that relationship. It is a very beautiful relationship. They see the sky as their Father, the Earth as their Mother, and the Aborigines are the children.

The dream time that they speak of, is that part of every day for them or is that something that occurs once in a lifetime? How does that work in their culture?

Thank you for refocusing the question. It is something that every Aborigine does at least once in a lifetime. Many have multiple dream times. When they focus to meditate, they become receptive to the many things around them. These are often invisible to the eyes, but not invisible to the soul.

I think the American Indians also make that kind of connection.

Yes. It is the same behavior.

I have heard or read about the Count of St. Germaine having a relationship with or affinity for the Aborigines. He was a European who supposedly lived hundreds of years. He never appeared to age.

I am searching. I do not know of this relationship.

I do not know if it is true, but, if so, it is an interesting relationship.

Yes. It is always interesting. I will have to do my homework on that. My mind has access to many things, but I will answer you later.

Zeta Reticuli History

What is the furthest light spectrum that you can see?

We can see further into the infrared and ultraviolet, and also in the visual light range. Our eyes are such that they are larger and can gather more light. Our 'night vision,' as you might regard it, is excellent. It is useful in our many fieldtrips to your planet. We often appear in the dark to use our biological advantage over human eyes. In being able to conceal ourselves, we do not like to make our presence known. We like to leave a zero-residual presence. Many other species also share this. We can see in the ultraviolet and into the infrared. We can see body temperatures, if we learn to train ourselves. Our young learn this at a very young age, and then it becomes instinctual.

I imagine that, because you can see in these additional spectrums, you see life forms that humans are not aware of at all.

We see forms and also energies that can surround a life form. The energies that are radiated by life forms, sometimes called auras, are more vivid and distinct. These also contribute to the status of the life form, their intent, and their general attitude. We can tell whether they are a negative or positive form.

Are you talking about their energy signature or their color signature?

Both. The energy signature will sometimes be a peripheral layer within a color, a visual mark. It provides a greater depth of where the life form is, whether it is under stress, or if it is exerting energy, or if it is in a relaxed, static state.

I know that you 'vocalize' through thoughts and are able to vocalize through Steve. Do you have vocal cords that are able to emit sound or have you evolved beyond that?

We do. They have regressed and become less exercised.

But they are still viable if you need them?

Yes. We would sound like a crying animal, relative to humans. We use Steve's and Mary's voices to express in an audible, eloquent fashion. Most of our communication is through thought. We use verbal sounds. This cannot be ignored. It is part of what makes us who we are. Our vocal cords are a less exercised organ now, much like our hearing has diminished over time. Our mental capacity, our eyesight, and other body senses have compensated and increased. Our verbalization and hearing have decreased over time.

Spirituality

What do you anticipate might be a major shift concerning your use of telepathy? For example, at one time a major shift was going from consensus to neutrality.

(Answer not recorded due to disconnection on recorder.)

I apologize, Han. I had a disconnect.

I will repeat. I understand your technology.

(Therapist laughs about the foibles of the recording equipment.)

It is for us to understand this for many reasons. In this case, to explain, usually we are visiting many places, and we often deactivate these devices so that our presence is not detected. In this case, it is just the opposite. I will repeat. The telepathic range of our species allows us to see into the last moments of life, as well as seeing into the experiences of new life, from the earliest moments to the last moments. As a result, we have a broad and pervasive understanding in our culture that the afterlife is nothing to be feared. We feel that our next stage in telepathic evolution will be stronger connections with the afterlife. This is an ability that, in crossing that dimension, we will enhance next.

That brings forth another question about that. We talked about how a short life span tends to speed up evolution. Have the Zetas discussed the possibility of shortening their life spans so that more souls may come in and evolution speed up to reach the next phase?

I understand your question. If one is to regard the afterlife as the ultimate goal, that would be a logical stance. Also, we, to a degree, live for the moment and let nature take its course and enjoy the blessings that we have in this existence in our many levels of dimensions. We know that, in time, this will eventually transition to a spiritual one. Each is an individual journey, not in an evolutionary, social, cultural sense, but within the journey of the individual life form. We will eventually go to this stage quickly. Our society and its biological abilities will leap across that threshold in a physical sense. We already know that this can be done, but we must relinquish our physical bodies in order to facilitate this. We hope to broaden the perspective, but this will take time. In a sense, that dimensional jump or transition at the end of life is one that every life form will make at the end of a natural life span. Culturally, a whole society reaching that level may require an understanding that can only be related through each individual experience, and that accumulative effect will create a sense of connection with that afterlife dimension. The expectation is already in place. The ties that could be developed with the energies and abilities that conceivably exist to move and create without instrumentalities would require much more time. Whether we can evolve and enjoy those same abilities in our own physical place is the goal. Whether it can be achieved is rather problematic. I hope my answer is not confusing.

No. You come across with perfect logic and clarity.

For some humans it would seem frightening, illogical and dangerous. It is merely a more sophisticated view of reality, a mature one. Through our senses, we know the afterlife exists. It is only natural to be curious, explore it and to incorporate it into our own physical existence. Our beliefs lack the religious context.

Contact

Remember last session when we were discussing the word 'species' and how it is no longer appropriate in referring to your clan and the Earth clan?

Yes.

I have come up with a few ideas, and if you have some ideas, I would like to hear them. One thought I had was the term 'kindred spirit.' I thought we might take a part of it, and perhaps you could refer to Earthlings as Earth kin, and I could refer to your beings as 'Zeta kin.' I could use the name of your home planet, but you are all over Zeta Reticuli.

We use our origin planet. This is how we are associated.

And what should we call this?

Reticuli System.

It is a little long.

Yes, it is.

Could we shorten it a little bit? What would you suggest?

What was your first suggestion?

'Zeta kin,' referring to your Reticuli System inhabitants.

That would be appropriate. Zeta kin.

To me, it has a warmer sense.

I like that. Do you have any others?

I had thought of referring to individuals as 'kindred,' making it a noun rather than an adjective. I don't know if that would be necessary to use, but I am always open to inventing new words to express ideas.

I like Zeta kin. It is short and distinctive.

The expressions would only differentiate us by origin.

Zeta kin will refer to which?

Your origins and people. Earth kin would be my people. The kin part of the expression makes us almost family like.

The last part of the expression is connecting all of us. I commend you. That is what we will use, if that is all right with you.

Yes.

Thank you. Zeta kin and the Earth kin.

The Universe

Remember how you had discussed a certain type of energy, the spiral, serving as an anchoring device, being like the turning of the corn meal in a bowl?

Yes.

I heard a scientist speak last night, and I think it might be the same thing. He referred to it as torus or a form of energy that has a donut hole in the middle. He said the Universe mimics that particular shape. He describes it as a toroidal shape with vector equilibrium. It brought to mind your description.

Yes. That would be consistent.

He also mentioned that a number of contactees have described a similar energy flow in relationship to propulsion systems in UFOs.

This is a common technique.

There was also something called the Flower of Egypt, a symbol found on walls in Egypt, that they now believe is a two-dimension diagram of a multi-dimensional system. Also, Tesla's work included this. The scientist said that entropy did not apply in this case. Entropy occurs in a closed system, but the toroidal shape, or the one you described, is an open system.

This is true.

It was pleasing to me to find that little bit of additional information.

There are external forces that are required to propel these phenomena that can only exist in an open system, hence the lack of entropy, which exists in a closed system.

Other

Is Steve doing well with his project and his helpers? (This is a reference to the project assigned to Steve by the therapist to keep his mind occupied while Han uses Steve's body to telepathically communicate without Steve's awareness.)

Yes. He is studying the necessary changes to the temperature and atmosphere of the planet and how these changes are going to be executed by using appropriate energies and matter. He is also studying the basic change in gravitational forces in the atmosphere and determining the temperatures that are necessary or what must be altered to form the basis for changing the atmosphere to one that is compatible to many species, including humans. *(Steve's assignment is to mentally terraform a planet, using off-planet assistants.)*

You know that I sent him 'helpers,' and I do not know if these helpers are being supplied by the universal consciousness, his own unconscious mind, or if he is getting help from your Zeta kin. This is to give him a chance to stretch.

Helpers Scala and Lent are an 'experience' and are educating Steve, along with Ms. Peters. Ms. Peters is the name of the being that greets him and also takes care of important tasks. There is much to comprehend, and this is a phase of stretching and learning to understand the scope of what is to be done. In particular, this is the preliminary work to achieve a certain end, using terraforming or planetary engineering. Each planet will create its own outcome naturally, anyway. The job is to take a planet where life already exists and match the setting for life while terraforming, so that the existing life can continue. It is the opposite of many natural processes where the natural conditions are established first and life evolves as a result of that. With planetary engineering, it is the opposite of that.

Open Session for Comments and Questions

Is there anything else that you would like to discuss or offer before I bring Steve back?

I have spoken of others who visit your world who are not always welcome. We could harm them in a very terrible way, but our moral level does not permit that in ourselves. Instead, we work with them in a peaceful way to persuade them to act differently. These behavior changes are not always immediate and not uniformly accepted.

If we wanted to make a social change, we would work with the young.

Begin at the early impressions stage. These impressions and early development extend for the rest of their lives. It is there that change is facilitated in a personal way. It is the only way that humans truly change and should change.

With that and lots of good thoughts for you and your kin, I bid you a fond farewell, until we are able to talk again.

Steve has a capacity only for a limited time.

Yes. He is getting tired.

We both understand this. I could and wish that we could talk longer, but I understand the limitations and accept them. Steve is tired, and yet I wish to talk more with you. I understand.

Know that our dreams are open to you.

I look forward to the day when we can talk for a very long time on a park bench among the stars.

And have lots of laughter exchanged, too.

Yes. As we discussed previously, I thank you for the new terms of 'Zeta kin' and 'Earth kin,' and soon our kinship will unite when we meet again. Thank you and goodbye.

Session Thirty-Four
March 11, 2012

Client is inducted. For the third time, therapist hypnotically removed Steve, the client, from the exchange between the therapist and Han by creating another scenario in the client's mind for him to pursue. It was suggested to him that he was now on a bullet train, entering a tunnel, and going to work at 232 Main Street, 4th Floor, in a mythical city. In this scenario, the client was given a large office with a conference table. This therapist assigned Steve the task of preparing a new planet for habitation by developing plans for terraforming a planet for new life forms. The plans were detailed and included changes to the atmosphere, water, Earth, and new plant forms. Client was 'given' two assistants, Scala and Lent. Steve checked in with the receptionist, Ms. Peters. (This is scenario creation technique to remove the client from conversations between Han and the therapist.)

Han, you are most cordially invited.

Hello, Mary. This is Han. I introduce Gen, who is with us today

Oh, excellent. Welcome Gen. Thank you for coming.

Gen: Hello. I have looked forward to meeting you.

Han: He is standing near me, but slightly behind me. I am sitting down now.

He is in the background so he does not encroach, and so I can focus. At the same time, he hears what is being said. He knows in a social situation there is a sense of space and boundaries. We know, in your case, that does not apply. Gen sits back so that social conversations can flow to the back and to the side. He is attentive to the sense of proportion in space boundaries in conversations for some. It is different for us, since we can be aware of the thoughts of others across a large room.

Yes. And much further I am sure.

Yes. We can 'hear' much further.

Earth History

I am curious about a behavior that I am sure extends beyond human beings. That is the behavior involving the masquerade. We even have masquerade balls. I am pretty sure that you have come across other life forms that like to play with hiding their identity in a playful way. Would you care to tell me about some of those experiences, and perhaps tell me why the individuals enjoy this behavior?

I have never understood the human behavior for this. It is part of your biology. You can limit perceptions and permit this social situation. Perhaps it is a means in a party situation to permit interaction with someone that one might not normally interact with, thus creating new viewpoints of those behind the mask. The perception is one-way from those seeing through the masks. The human person does not know who they are talking to. They can hear the voice. It is a curious situation. We could telepathically see through all of this and already know who they are. Knowing who one is becomes more important than knowing what one looks like.

It serves no purpose from your perspective?

We understand it, but it does not serve much purpose. It is a quaint children's game often played by adults.

I find it quaint myself and was curious why it would be considered playful.

To me, it would seem almost deceptive, and, depending upon the personality of those there, it could be considered threatening due to not knowing who is behind the mask. It requires a sense of emotional risk. Perhaps that is the whole purpose, but it seems rather quaint since we could see through that in a telepathic sense. We know that humans do not yet have that perception. I am scanning some of Steve's thoughts. When he was single, and in single groups, he would often feel like a fish out of water. In a sense, these masks place everyone in the room on the same level. I am sensing in Steve, the social awkwardness of it. Steve prefers seeing everything plainly and honestly. Often that is not the social role in a party where such masks prevent this. He can conceive of it, but he has never been to such a party.

When in a crowd, human beings may feel that they cannot be identified easily. When they feel their identify is concealed, they feel more at liberty to act out on any impulse they may have because they will not be found out.

It has a liberating effect, but it can also be taken to extremes that may not be appropriate. I can see that now. I understand. That is a perspective that I did not consider. By concealing and gaining a perceived anonymity, it releases human inhibitions.

Steve is happy somewhere else, but at a distance. I can read his thoughts. This human game of concealment would be a temporary situation that would need to be exposed eventually. *(Han was reading Steve's thoughts about his youthful social experiences that were sometimes fraught with concealment behaviors.)*

Yes. Under normal circumstances, that would be right.

I will not disturb you with his reply.

I want to keep him far away right now.

I sensed your thoughts. These contrivances *(Steve's temporarily induced daydreams)* do not show one's true self. At the same time, that is the purpose, to permit behavior that might seem more acceptable in the context. Much human behavior, as to what is appropriate or inappropriate, often changes. We have learned from studying your behavior, when humans agree upon an accepted behavior in a specified context, that it can change from situation to another.

Yes. That is true.

Did you have an opportunity to investigate the extraordinarily long life of the Count of St. Germaine and a possible connection with the Aborigines?

No. I have not.

Then we will skip the question.

I will refer back to you next week. I am sorry. I did not know I had homework.

What connection was there between Lake Titicaca and Easter Island, particularly in reference to a written language?

Lake Titicaca is a high-altitude lake. The water has some unique species. There are some unique linguistic patterns found in these two locations. When you posed the question, I smiled for a moment and almost thought of telling a joke, but.... *(thought left uncompleted).*

It would be good to hear a joke.

I was going to say that your Easter Island was scooped out of the Earth, turned upside down and placed in the ocean to make the island and the emptied space became Lake Titicaca. But we are not traversing space to do something like that. (Said with dry humor.)

(Therapist laughs.) I enjoy the thought.

It would be more a play on how you perceive us and what you would think we would be capable of doing. That is all. You probably think we could do that very easily. Perhaps we could, but why would we want to?

(Therapist continues to laugh.)

There was a linguistic connection. In the cultures that existed in both locations, there was a connection between them.

Were these cultures composed of off-planet beings, or were they Earth natives?

As humans study more of their past, they are looking at an album of their lives across thousands and thousands of years. There is an unconscious connection, an intuition, that there is more than first meets your eye. This is true in this situation. There are extraterrestrial connections in both locations. This is easy to say because there are connections all over the surface of the Earth. These two locations are not too different from this. They do not distinguish themselves more or less than any other places we have visited. These are cultures that we have interacted with in the very distant past. The attention and capacity for human memory has faded to an extent that it naturally becomes a form of concealment of our past activities with you.

That is probably pretty easy to do.

We are often regarded as deities in these situations because we are different in appearance, yet similar. They immediately place the cloak of a deity around us. We understand this human eagerness to do this. It is awkward for us, but we can understand why something appearing more advanced would cause this. We have traveled enough to know that what one perceives as 'advanced' is not necessarily the case. One can be ecologically advanced, but not morally advanced. With respect, we know that a little humility is required, and that is why we do not exploit this belief in some of your cultures. Nevertheless, it exists. We have tried to help in matters in the architecture and engineering of their structures and provided modern ways in sewage systems that were very sophisticated for their time. We tried to provide some basic conveniences to make them a successful culture by providing help in their infrastructure. Some of this was done by creating in the minds of the inhabitants, ideas and knowledge, teaching them these things.

Certain concepts can only be fulfilled with a certain kind of energy or advanced capacities which, at that point, we provided them. We instilled in them ways to join rocks with very tight tolerances, interconnecting components, which created strong structural dwellings. We used the basic materials provided so that we did not betray our presence in the future with more advanced civilizations, having an interpretation of old accepted thoughts. So, yes, we have interacted with these Mayan and Easter Island cultures. Their survival was one that did not often follow some of our precepts. They often defoliated their forests. They did not understand. They lived more for the immediate instead of for long-term survival. Much of their forests were eliminated, which eventually came back to harm them, and they eventually disappeared from the island. They did not understand the sense of environmental balance and living in a positive way. The residents at Lake Titicaca, on the other hand, were very successful in each location situation. Here we have two contrasting cultures with very similar origins. Yet in one case, they destroyed their environment, and they disappeared, leaving behind only many enigmatic images and remains. In contrast, the high mountain residents, with their unique environment, successfully adapted.

Concerning crystals or stones, do they have any energy capabilities to change health or mood. Is the application of gemstone energy to human beings of any use?

Not in a direct physical sense. Crystals do have energy beyond their light reflecting and refracting abilities. It is the belief that they have certain abilities that heal. They just trigger within the individual person his or her own capacities for healing. The gems are merely activating triggers. It is the human healing capacity itself that does the healing.

Is the trigger activated by belief that the stone is having an effect?

Yes. The human is motivated by a belief. It is belief that makes the individuals aware that they have certain energies within themselves. This also comes from the external, that which surrounds them. At that point, they become conscious of this, and they harness or focus this energy, which helps them. The energies come from another place, but the crystals help them focus.

It sounds a little like when I work with healing, I pull energy from the Cosmos.

Yes.

Spirituality

Are you familiar with what we call 'mediums?'

Yes. I believe so. It is those who talk to what we call the deceased. We are doing similar things here, though I am not 'deceased,' yet.

(Therapist laughs heartily.) I am so glad.

But the technique is similar.

Yes. There is a particular thing that happens with some mediums when they are sitting in a deep 'trance.' Sometimes a material called ectoplasm will pour out of their orifices and float a little bit. I really don't know what it is. It has been photographed. Do you have any ideas on that?

It may be from the 'medium' or spirit from the beyond, and it is transitioning over to this reality. It is part of that inter-dimensional flux between their reality in the afterlife and this existence. That is my only limited understanding.

Well. I learned a little bit from it. Thank you for that. It is a flux.

It is a byproduct of the interface between the living (physical) entity and the spirit from another existence. Due to our telepathy, we can experience the instant of death and the moments that approach death. We can also feel a great sense of energy and release for a brief time after. This energy release gives us solace and encouragement that death is not to be feared. Beyond that, we, too, just have a temporary deeper glimpse. Beyond that, death is a mystery for us, also.

Do you physically interact with that releasing energy when someone dies nearby you?

Yes. There is joy and peacefulness. It is quite inspirational and warming to our souls. It emphasizes to us that it is not a loss to the one who dies, but it is only a loss to those of us who remain. Sometimes a parting is like a burst of energy, like a light bulb filament, which burns brightest before it burns out. It is not ever burning out. It is just going to another place. We can experience this to a degree. It touches our souls, and it has influenced us and affected our culture, as it has in your culture, in how we perceive things.

With the Zeta kin, if one were to transition near you, do you get the impression that the intelligence released is much larger and much more beautiful than when it was contained?

Yes. It is.

I have also had that experience.

Yes. It is like pouring water out of a vase. Strange things happen when the body pours out a soul that could fill a swimming pool.

That's a good analogy.

It confounds physical concepts, but not spiritual ones. The body, whatever its purpose, is a confining container. Once the spirit is released, it takes on a larger, more dispersed presence with a strong sense of consciousness and identity. It has a larger footprint. In the hereafter, there is much spaciousness there to receive such a spacious consciousness. It is like a cocoon, and the butterfly is released. It is a much larger object, with more color and freedom. It has given us happiness for the ones who are close to us to see them in that state and then, for a brief time, several minutes, we see them in their state before they move on. They slowly fade. We know that they do not just disappear, but are moving on to another dimension. It is an invisible door that we cannot see, and we cannot enter at the present time. Death has a meaning to us that it is not to be feared. We do not diminish the value of our current life. We know we need to conduct ourselves here for the role given to us, and there will be plenty of time for the next role that comes in spirit. As we grow older, our physical bodies will feel a dramatic change, and we will be released, unfettered.

Unfettered is an excellent word. Since you have experienced spirit transitioning out of the physical form, did you experience an emotional component in their energy?

Yes. There is a sense of joy and a sense of such radiance. We know the capacity lies within us. It is like the body has been bound, concealing parts of ourselves. Perhaps it is our society that has felt the need to protect itself by repressing this. Yet we know that it will compensate in other ways, if one wishes to be unbound. At the point of life's end and transition, in one sense, the full flower of a soul has been released from its container. Philosophically, we see what we, in this physical society, would strive for. We see great beauty in the identity and consciousness (of the soul) and we see certain things we admire that we see diminished in us. These things finally become unrestrained in death. Our society and culture need to focus on increasing those areas because we can see the soul for what it is, for that brief moment. We need to try to emulate that all of the other time.

Yes. It is stunning to see that. It is a worthy goal to experience that all of the rest of the time. I see our physical forms, to use the analogy you used last time, as rather like flutes. With each form we take, there are only certain holes available to be played. I don't know, but perhaps if we take different forms, we are able to play different notes.

This is true. We see the full range of greater music for that brief moment as the soul transitions. It is a lot to look forward to and, at the same time, it gives us pause to consider what is inside of ourselves, and what we can strive for it. We can try to live in a way that will not only enrich society, but will also enrich the individual with its capacity for joy and happiness in intellectual pursuits, all in proper proportion. I have talked about our society as a community of mind. Here we are talking about the individual, spiritual sense.

Steve had a dream this morning in which Joe, his father-in-law, who recently passed away, was giving a lecture to a full audience. He was describing his experiences to the audience and telling what it was like since he passed over. Do you think this was a visitation dream, or was it simply a dream where Steve was making a connection in his own mind?

It was both. It was a connection Steve was making, reaching out and, at the same time, it was a response, a presence that came and reassured him. The images that Steve has in his mind are vivid. Uniquely, the room was a terraced room, like a lecture hall, like a theatre. Interestingly, the speaker, Steve's father-in-law, was facing in the same direction as the audience. They faced a big projected image, and Joe's thoughts and impressions were being displayed. In Steve's mind, what was on the screen were pleasant images, but no specific images remained in his mind. There was an informality. Many of the chairs closest to Joe contained his family and loved ones, who were listening intently, sometimes reaching out and touching him. Around these informal folding chairs was a more formal terraced row of people listening. Those closest to Joe were not so much watching what was on the screen that Joe was eager to present, but they were just mesmerized by his presence and were enjoying his energy that was comforting to them. Steve's wife was particularly comforted, as was her sister. They were nearby, touching and embracing him. This did not deter their father from giving his lecture, as he had often done in life. His presentation was like the slides one might show of one's vacation, only this was the afterlife. It was a happy time. It was one that Steve felt he needed. He felt appreciative of his father-in-law. This enabled Steve to visualize what might occur with his parents in the afterlife. It was a sense of comforting reassurance. It came to him out of nowhere, but he partially reached out. He did not expect such a clear vision to be placed in him. Of what we know of the experience of some mediums, in the spirit world, spirits are not good at keeping appointments.

(Therapist laughs.)

Perhaps this is as it should be. In this case, it was most welcome.

Do you know if the dream was shared with others or was it just sent to Steve?

Steve shared this dream with his wife. It is uncertain, though, whether it had the same happy meaning for her. He wondered if she understood what it meant to him.

She did not participate in the dream?

This dream was just for Steve. He has the capacity to be more open-minded and, as such, he is more receptive. His spouse has certain restricted views of reality. Perhaps they are just as good, but they are more formal. Steve has a bigger, more open-minded capacity to accept contacts in ways that are not conventional. He has no preconceived notions.

I find it useful that his mind is that way. It helps to facilitate this process.

It was a visitation, but it was also a response to an invitation that he had extended for some time.

I think he finds it personally reassuring.

Steve is sometimes lonely, even in a crowded room. He has some instinctual capacity to know that there is something beyond his own reality, but he doesn't know quite what it is. He just knows that it is there. That gives him some comfort and, at the same time, stimulates his curiosity. His dream today has given him some peace, and his household is quiet and calm today. It is not always the case. It is a peaceful Sunday.

Contact

As you know, Steve gets a little nervous when you extract information from his mind and use it, along with other sources, to respond to a question. He is very concerned because he does not want to be the contributor. Please kindly explain to him what resources you use and how you extract data.

I pluck thoughts from his mind and others to create the pattern. It is like I am making a sandwich. I pluck a little here and a little there from the 'deli' of thoughts and make a 'sandwich' appropriate for explaining what I wish to convey. In his humility, he sometimes feels that his thoughts are not of much use and, from his perceptive, an advanced species would not use his 'weird' thoughts when they must have such greater capacity. However, in order to interact, it is necessary for us to perceive the thoughts of those we interact with. He has often felt discouraged and ignored by many, either because they do not understand or his ideas are not acknowledged, particularly in the past. He should not feel this way. Some others have other agendas that are perhaps not as magnanimous as his. His thoughts of others are not often given credence. Others will sometimes dominate his thoughts. His thoughts and his desire to help others have value. This is what I sometimes sense in him. I also pluck thoughts from others. There is no sense of intrusion. No sense of privacy is violated. The thoughts are, from our perspective, merely out in the open, easily accessed. Nothing is taken away. One is only illuminated by what is thought. The sense of modesty or privacy that humans have is rather unique to them. Part of this is due to their lack of telepathic ability.

Would you describe the human mind as simply being a kind of library from which you can gather information? Perhaps it is not as organized as a library?

The thoughts from some humans can be well ordered. Most are not. Sometimes there is, between genders, a human capacity to kid the other gender about one having more ordered thoughts than the other. This is irrelevant, and I can speak with authority. I am able to read these thoughts. The female and male thought patterns are often different. However, overall, they are very much the same. The gender is rather inconsequential. In general, many thoughts are not ordered when the situation is not understood. The thoughts of confusion and fear exist in several different ways. The mind is the repository of many thoughts, senses and impressions. There is also, in some, a degree of chemical or mental imbalance, which we accept. In their diminished capacity to understand their reality, they have been further confined, only responding on certain instinctual levels. We accept this. We have no less respect for them than for others with greater capacities. Each is unique unto themselves and are accepted as such.

Yes. I have learned to respect and enjoy 'special' children and their unique minds.

Human children, in that state, at that age, often have such great capacity for learning. It would be a human goal to impart the wisdom from long experience and age and place it into the mind of a young child. Sometimes, when humans age, they can revert to a child-like state. It often includes diminished mental capacity. It is a downside. The ideal situation is the mind filled with child-like curiosity, coupled with the wisdom, knowledge and experience that old age provides. That would be a worthwhile goal, if possible. Such is life.

Other

When you feel emotion, do you feel it only through the thought process, or is it a physical reverberation?

Both.

And the reverberation, does that include the entire body, or is it specific to one area?

There is a specific area. Our thought processes predominate, but our conscious and unconscious receptors for feeling exist separately. They coincide with thought. Thought can only interpret a situation analytically. Emotionally, there is another part of us, perhaps more limited, that will struggle to keep up. In that area, with humility, perhaps I have shown a great emotional affection. It is only to the biological limit of my species. Perhaps it is sufficient. Within our psychology, compared to humans, I sometimes feel we are diminished in that capacity. Our emotional range is like our hearing. It is reduced over our evolution. I am not ashamed of our biology. I am just pointing out certain differences between Earth kin vs. the Zeta kin, but we probably have a greater capacity than we realize. It has been something that we have suppressed in our evolutionary past for our own wellbeing and peace for the good of many because our sense of reality is so connected to others in thought and being through the eyes of others.

I think that creates true intimacy.

This is probably true. It is not common for an advanced species to share such vulnerabilities. I suppose it is part of the maturity that we share in our exchange and relationship that I can share that. We are different, yet we are so much the same. Our evolution has deemed upon itself to exist in a way that is best and appropriate for our species. Maybe your species will evolve in a similar manner. We feel an unconscious connection with humans and wish to reinvigorate our emotional profile. It is not, in our minds, without danger, and we have a fear of reawakening the mindless primitive that exists in all living creatures as they evolve away from primitive times to the present. This is a broader perspective, a connection with others of our species.

As you concentrate and reintroduce yourselves to emotional experiences, can you not confine it to the positive emotions?

That is our focus. That is our goal. Though it may include sadness, and we understand it, we must focus on what many consider the positive energies.

I pick up your intentions and your sense of caring in a very large sense.

I understand. That is the irony. We are probably capable of more when we think with our telepathic abilities that many humans would envy. It has helped us survive and thrive. These are some advantages. We sometimes wonder what that capacity has displaced within each of our souls that we feel is also important to fully experience.

Open Session for Comments and Questions

Does Gen want to make any contributions?

He is enjoying the sky, and he thinks you are lovely. He feels that the purpose here is good and extraordinary. I feel boundless energy from him, yet there is a calm, and he senses my mood. His excitement is reflected in a way that it would not interfere with our discussions. I sense this seashore with mountains, and he is sitting on a beach, observing the sky with a band of light, the Milky Way. He is enjoying the cool night air. He is examining flora and the animals. He can often do this in a way that he seems invisible to them. He can shift dimensionally so that he might seem to disappear, and yet be present. He has never been to Earth before, and I think he wanted to come to see and enjoy this, even though he has been on many other worlds. He is enjoying the beauty of your planet, and he has now returned from the beach and is near me, being attentive. He has returned from his exploring and wishes to do whatever you wish to do at this point.

I wish to thank him for coming and visiting.

He thanks you, and he thanks you for the friendship with me (Han).

Han, do you have any parting remarks? My dog is getting hungry.

Oh. I can see your dog. She is penetrating, and her tail is wagging. She is nuzzling close to get your attention. I see this.

I hope Steve worked very hard and accomplished something. (This refers to the side journey he takes during hypnotic sessions.)

Steve has left images of his traveling through the tunnel. When he accelerates leaving the station, the whine from the train increases in pitch and then fades at the same time, like a sound becoming more distant. Then the train disappears around him, and he is no longer sitting in a seat, but in empty space with the stars surrounding him. As he reaches his destination, his immediate surroundings begin to fade back in. Then, before he knows it, he is in a station again. This traveling phases through time/space and then, magically, he leaves the train and goes to his office. He is addressing and developing a number of complex relationships in temperature and weather for the new planet. These relationships are now being resolved, which will provide the basic fundamentals of a beautiful garden. It looks very rough right now in the mind in terms of what needs to be done.

Yes. I will be asking him to share that when I bring him back. I don't have sufficient language to express my gratitude to you for being willing to share your thoughts with us and to exchange ideas. I think you know that Steve and I are very grateful.

We appreciate that and are flattered that you are so receptive. We also appreciate you. Sometimes my hand reaches out to touch you on the shoulder, as is our custom. I have, what to some would call, spindly fingers, but there is a gentle touch. It is a touch that wishes to express our appreciation for you and your like-minded friends. There are many human beliefs and concepts about how extraterrestrial contact would be made. It is not generally understood by humans that this (telepathic connection) is often how such contacts begin and flourish from that point. This is usually how it happens, and it is the normal way. In a sense, it is you who are the extraterrestrial, and yet we know that you regard us in kind. We are all kin of the Cosmos.

I am just extraordinarily happy about this.

We are not the Hollywood stereotype as you might have expected us to be. This is reality. It is not fantasy, and it is the means that we are using to communicate today.

Thank you and, until we meet again, Namaste.

Namaste, until we meet again. I look forward to our next meeting next week. Farewell for now. For Gen and myself, we will do our part before we return again for our next conversation on the park bench of the Cosmos.

Session Thirty-Five
March 18, 2012

Client was inducted and mentally taken to his worksite where he is developing plans for terraforming a planet with his assistants Scala and Lent. Upon entering his office, he is asked to describe who is present.

Steve: There are two. One is near my desk, reading, then inspecting me, and the other one is in the office in another area, also reading, but attending to something else and is now putting things down on the table and walking over to greet me. I walk over to them and greet them with a 'Good morning.' We are now sitting around the table. *(Therapist then gives directions to Steve and the others.)*

There is an agenda prepared for this meeting. It is there in front of you. All of you will be working from it today. I leave you temporarily to your work. I am always available. You are always safe. Work with enthusiasm and enjoy it, and I will see you later. (Therapist leaves Steve to mentally engage in his assigned terraforming task and shifts her attention to Han.)

Han, you are cordially invited to speak with us, along with any others you care to bring with you.

This is Han. There is another colleague with me.

A different one (Gen) from last time?

This is Zestra. She is a female.

I welcome her and anyone else that you wish to bring at any time.

She may appear to be without gender, like me, but from what she projects, she is female within her. She is a colleague, as were the others who have attended in the past. They were males. This one is female. The gender is of no other consequence.

Has she been to this particular planet before?

Yes. She has. She is attending this time. She is impressing you, greeting you, imprinting you with thoughts of welcome.

Thank you very much.

She is standing back now, relatively passive and aware of her surroundings. She is primarily communicating with you through me. She is also free to communicate, should she choose to do so, just like the males who have visited in the past, too.

I would welcome that.

She understands and appreciates this. She does not look much different from me. I have never presented a female of my species. We appear much the same, but in our souls, there is a gender difference, and often there is a maternal or feminine or a masculine quality. Often the two aspects are very much merged. To a human it may not be too distinguishing, but within our species, this is what is considered quite normal for us.

Zestra, you might enjoy my next question then, about a Dr. John Gray who has spent many years discussing and writing books about the differences between men and women. I find much of it humorous. He feels that we have become too much alike. He has written some information about the condition called ADHD and has recommended that Earth-based beings take something called grape-seed extract and vitamin C in certain combinations to level out the symptoms of ADHD. It might actually be the extract resveratrol. Would either of you care to comment on whether you think that combination might be beneficial for that condition?

This is related to a human condition regarding ADHD?

Yes. That is correct.

It would be a nutrient treatment for that condition. I do not have much information. Zestra is speaking now, and speaking for her, she says that this would be a partial treatment for that condition, but it would need higher dosage to have more effect. Dr. Gray's theory about the treatment would be successful, in her opinion.

Thank you very much for that.

Notice that, compared to my male companions, Zestra does not mind speaking out. She wishes to assist to provide you with more of an answer.

Thank you. Every little bit helps.

She is not holding back from you.

Earth History

This question is about the human brain. Who or what is using or influencing portions of our brains? I understand there are other influences.

Other entities?

Entities, yes. The influences could be interdimensional, too.

There are influences, interdimensional and spiritual ones that often affect parts of the human mind and influence it. The human, in understanding its reality, is often unaware of these influences. We have accepted this. This is partly due to knowledge and understanding of ourselves and of these other dimensions, with which you are not yet well acquainted. It will round out your understanding of your reality. There are influences beyond what your reality is willing to accept at this time. I have often spoken of the species that we are impressed with, which does not have a physical body, but is more or less pure energy, with its own consciousness. We are aware of this. In our interaction with them, we are learning to develop parts of our own brains to understand and accept this and to interact with them.

I am sure the human mind will, in time, have similar challenges, which I am sure it will accept and gain understanding. There are spiritual energies and forces. In our early sessions, we spoke of resonance, resonating energy that connects all things and all life. In time, humans will realize this connection and will more fully integrate into that resonating energy that permeates all life and physical objects in the Universe.

Does that energy have a specific hertz?

A specific....?

A specific measurable vibration, if that is the right term?

The speed is the frequency. It is the song of the Universe. It is the tone, the hum of reality, and the connection of all that exists in our reality together. I am speaking in a philosophical way. Forgive me. There is a frequency harmony, and Zestra is insisting that there is a tonal frequency that is more listened to by the soul than in a physical way, yet it is just as real. She adds that this is, in one sense, a truer reality that permits the communication in a quantum way across great distances, as one example of this.

That is why it is not realistic to try to interpret it on this physical plane. I do appreciate her response on that.

This is correct. It is not to be interpreted in a physical way. Yet it is just as real as the physical.

Well, it certainly is for me.

Are you familiar with the Georgia Guide Stones made from granite? The entire area is given to granite quarries. The stone monuments were built outside of a small town in the State of Georgia in a semi-wilderness area, and they are inscribed in eight languages. They list principles for human behavior and interaction. The first principle states that the human population should be kept below 500,000,000. I suspect that there is more than one set of stones around the planet. Could you give us some insights about these stones?

Some species, due to their self-image, often like to leave special markers behind. You could call it pride in their species, or they feel they must contribute wherever they travel by leaving a mark behind. Some species have visited the Earth and left a sense of how the Earth population should be fashioned. These create spiritual and cultural beliefs. Many early cultures have reflected this by worshiping beings from beyond, or from the stars, or other dimensions. These experiences are often reflected in stone as a reminder. Due to the past incapacity of humans to preserve their history, humans often created monuments out of stone to remind them across time of these visitors. Some of these exist today, but are not recognized as such because some modern interpretations do not accept that many space travelers could visit here. That is often the case.

An individual went to a banker and asked the banker to employ a contractor to have these moments built because there were good stone cutters in the area. At least that is the story.

These stones were created recently?

They were erected 1979-1980. The implication to me was that the Earth would suffer severe catastrophes, and that the remaining mankind might need something like the ten commandments as a guide.

These would be moral touchstones to continue by. The Earth kin seem preoccupied by their own self destruction. In light of how it sees itself, that view might be warranted, if one is to take a pessimistic view. This should be counterbalanced by potential for positive outcome as well. It is part of the psychology of your species to feel that, perhaps, it is too late. You are influenced by many external things from off-planet influences and internally, with spiritual and other resonating forces. Still, to a certain extent, you are the masters of your own fate. My caution is for you to be careful about what you wish for your future.

Thank you. When you reminded me that the bacteria that created the dinosaurs still exists and, therefore, the potential for rebirth exists, it had the effect upon me to not be so concerned about species becoming extinct because the elements to recreate them continues on. I am not a great worrier.

There would be different shapes from the same clay.

Yes.

In a very simplistic way, it is the same on other planets as well.

One of the things that I know will control population is that the viruses keep evolving and changing. I am a little concerned that some laboratories have taken a virus, the swine flu virus, or the bird flu virus, and have modified it to make it deadlier in the hope of coming up with a vaccine. I find that to be a very naïve and dangerous thing to do.

It is.

What could you tell us about viruses? I know, except for one or two cases, there is nothing to keep them at bay, and perhaps attempts to stop them only make it worse. I would very much like to hear what you have to say about viruses.

Viruses are constantly mutating and changing. In rare cases they can mutate into something that the human species would have difficulty coping with medically. Occasionally, humanity has come close to a precipice, and, as quickly as it mutates into a dangerous form, it can mutate again into a non-dangerous form. Sometimes human manipulation of viruses is done for dangerous and narrow reasons. Sometimes it is done to create a weapon. Sometimes it is done with the best of intentions to prevent that day when they fear they are no longer able to medically cope with a viral crisis. Caution and wisdom should be practiced long-term in making such decisions. Without full consideration, these decisions could be dangerous by modifying some viruses beyond what already occurs naturally in anticipation of protecting the human species.

What is the time span of a virus mutation? I know with the flu viruses, it seems like they change every year.

This is true.

When people get a flu vaccination this year, are they are already using something created last year, which may be obsolete?

Many of the previous vaccines form the basis for the new strains. You have to anticipate one of several kinds of mutations that are possible. It takes insight and skill to anticipate a new virus form and create a vaccine to combat the virus.

I don't think we have that skill well refined.

You attempt to do so every year, more or less successfully. The time may come when it may become more challenging.

It sounds like you are telling me that viruses, by their very nature, change all of the time. Even if one were to develop a weaponized strain, the virus would ultimately change. Is that correct?

This is correct. It does change. Usually, small isolated attempts to weaponize a virus are not practical.

I am glad to hear that.

Mostly, this is what occurs naturally anyway. Often the danger is that a virus mutates into something that is beyond the ability of Earth medicine to manage. This may cause a drastic reduction in your population. Without wishing to sound cold, to use your human term, this would not necessarily be unbeneficial. The Earth population is too great.

Isn't that simply the way of nature for populations to rise and fall?

Yes. It is forestalled by human knowledge to cherish life and to prevent death. Many in medicine are dedicated to protect and preserve human life. Yet it only forestalls the inevitable.

I don't see that as necessarily negative. Disease has its own purpose.

Spirituality

I know a minister who is also a medium, and she sees, among other forms of life, very small, playful forms of life that she calls 'joy guides.' I don't know much about this, but some may be called pixies or gnomes.

Fairies?

Perhaps. I refer just to smaller energy forms. Is that a universal concept that there are very small life forms that like to interact with other beings?

There are many manners of energies that exist, like many species of physical beings. Pixies would be considered one. Some might consider them playful, but they may also be serious, too, in their interaction with humans.

Some humans say that, with practice, they can sense them or see them. Perhaps that is what the minister is seeing, some wavering of light energy.

Contact

Now, if we can keep Steve far away and unaware, he wonders if he has only been visited once, aside from our communication, or if he has had other visits in his childhood.

He was contacted early and was being prepared for the future. However, at a point, prior to his chronic disease, he was no longer necessary for that preparation, so the contact stopped. However, the echo of those times has not diminished, and he is mentally very perceptive. This is one benefit from those times that continues to this day. We were saddened by his condition, but we did not take remedying steps to prevent it. Still, his ability to be likeminded and receptive has continued. His early contacts were very positive. It explains his lack of fear toward these contacts that we make weekly. He is very comfortable and accepting in his role and is not afraid of the contact. There is a warmth and affection he has toward us, like an old friend. His role has now changed, but the earlier relationship, though not readily apparent, is not forgotten and has not changed. It will remain so as long as he lives. As a child, he was shown many things, some of which he does not recall. Some he does recall. I regret that these contacts were concealed with unpleasant memories based on frightening concepts that humans embrace in their Earth mythology. In hindsight, this was unnecessary to impose all those thoughts on a child. We are not perfect. We found out that this is not necessary. His tolerance and acceptance demonstrate that it was unnecessary, and it is characterized by his lack of fear toward these sessions. It is what makes our communication so open and informal. Through Steve, there is a closeness that is facilitated that enables the intimate communication in description and in detail that has been typical of our sessions.

It has helped him in so many ways.

Yes. He sees a larger universe now. It is expansive. His imagination is strengthened by the insights that he has, which, through me, I also share with you. In a sense, he finds peace now, and perhaps what he learned as a child is now enabling him to fulfill the role that was destined for him. His physical condition is not one that detracts from this ability to communicate here. In a sense, he is now performing the role expected of the young child. Perhaps these sessions have enabled him to fulfill his longing for contact that he felt when he was a young boy.

He is able to take the added insights and be more of assistance to those distracted persons who call him and bring them some peace.

He suffers many distractions. It is just that some around him are needy. He tolerates this and has accepted this in his past. He is stepping back to us now.

I think it provides an additional insight. He can see the neediness that others place upon him and can understand why the Zeta kin do not wish for Earthlings to be oppressively needy in their relationship with the Zeta kin.

Any healthy relationship must be one of relative equality. We know, in time, the Earth kin will reach a level where there will be a relationship without dependence on the other, one with equal give and take. This will happen. This is the only true relationship that can last to accomplish an important, long-term, specific goal.

Things are changing. I was at a wedding, and there were some seniors sitting around a table. One woman brought up the 'shocking' news that the History Channel was putting on a program about alien contact. She made the point that 'contact' was not possible, in her opinion. She was dismayed that the History Channel would even consider such a possibility. The gentleman sitting next to her said, 'Oh, I certainly see how that can be.' He was quite comfortable with that. So, there are some societal changes happening.

Contrast today with 1947 in your time. There have been changes in many areas. There is a gradual openness that comes from many areas. Some humans still think they are the center of the Universe, even if the planet they inhabit may no longer be considered to be. There are continued contacts with off-world species and your species. Excuse me, we now call your species 'kin.' I am sorry. I need to be reminded.

That is all right.

Zestra is pinching me in her thoughts to not do that. She is, as humans often say, kicking me under the table. This is my humor.

Your kin have changed and are more open. This is partly due to contact with other 'kin' from other worlds. Also, it is the knowledge of the Universe that has been exponentially increasing with your kin. What was once considered theory is now fact. You are now seeing planets, what you call exo-planets, and realizing that it (life) is more the norm than the exception. You are acclimating yourselves to the knowledge that contact is about to occur. In reality, like awakening from a dream, you have already been contacted. You are just coming out of a deep time. With your new awakening, you realize that you have had friends and contacts for some time. It is a good time.

I am so happy just to be here to enjoy this.

Contact will come in many forms, in ways that will not be recognized by Earth kin. Over time, certain understandings will be realized, and this is part of the normal process with the way the human mind works vs. our species. It sometimes is different. Anatomically our minds are different, but similar. As a result, from contact with many species, we know that there is a certain characteristic process of acceptance and tolerance. For each individual culture that we make contact with, each will eventually form a realization that they are not alone in the Universe. Actually, the Universe is quite crowded.

I sense that you know hundreds of thousands of cultures, some at your intellectual level, some not.

There is much variation in the many species in their individual evolutionary development. Most have similar origins. Some have different origins. You do not yet have the technology to enjoy interplanetary travel. When you do, you will have the opportunity to see the off-planet life. Your world became smaller in a cultural sense when you invented true space flight. This continues today. It is around the galaxy and around the Universe. The galaxies also become culturally smaller.

It is food for thought. May I talk to you about a contactee named Barbara who talked to Steve today for an hour or two?

Yes.

Her unusual blood type caught my attention, and her son's blood type, which is also unusual. She seemed to be telepathic, at least while she was aboard a craft. If you would care to comment on what these rare blood types mean that would be useful. I think she will be calling me, and I have to be careful about her feelings. If you care to share, it would be appreciated. She calls the tall ones 'Nordic,' but I don't know if that is the case.

She is apprehensive. She does not fully see her role as to why she is the way she is. She is a legacy. Due to interspecies development, she has some alien traits within her, as well as being quite human. She has certain abilities reflected in her blood type. The primary species in contact with her does so because of her interspecies biology.. She also received visits from others who wish to intrude upon her relationship with her prime parents. She has two sets of parents, her loving human parents and her silent kin reflected in her genetics. She has been unaware of this. She has often been afraid, and has coped with this all of her life. She does have many insights, but she still does not know why. It is because she is a legacy.

Her son, then, would also be a legacy?

Yes. Socially, he remains at home and is expected to do so all of his life. He has certain limitations and, at the same time, advanced abilities, as well. It is a combination of both. Some consider him deficient by human standards, while, at the same time, he has many advanced abilities as well, for example, computation.

It doesn't sound to me like she is ready yet to have a great deal of knowledge about this. To ameliorate this, when I work with people, I like to tell them stories of other people and let them draw their own conclusions.

Steve has often used analogies to communicate. It is a gentle, indirect route to the same message.

That is how humans take on and own information because they draw their own conclusions. This is effective.

Analogy is, to use another analogy, like window shopping at a store. They can observe what is being sold to them, and they are given the illusion of accepting or rejecting the 'goods.' In reality, the message has already been slipped into their pocket.

Exactly. I understand you.

The Universe

From an orbiting telescope from NASA's Solar Dynamics Observatory, March 2012, a photograph was taken of the sun. The image showed a black sphere with a tail going down into the sun's corona. The scientists called this tail an 'umbilical cord of plasma.' Do you have any ideas about what would be spherical and tapping the sun's corona?

An advanced species was tapping the sun's energy, an energy vortex. This was an undisclosed globed object, not a natural phenomenon, but one of very advanced technology. Someone was concealing themselves and tapping into the available source of energy of the sun for some undisclosed purpose.

Well, the photograph was certainly an unusual catch.

It was not seen before. It is one that has come to human awareness through their advanced technology. Humans may find they will observe this phenomenon more often now, and find that this is a more frequent event than humans have been imagined.

Other

You have probably observed over the years the huge political, religious, social, and emotional reaction to assisted suicide. I don't know that taking one's life in your culture would even be an issue because I would think it would be well thought through, and the individual would have the right to make that decision. Is that more or less the way the Zeta kin feel about that?

Sometimes we do suffer physical injuries which would, from a human perspective, cause one to consider assisted suicide. If accepted, it is accelerating the transition. Some would morally argue, prematurely, for the natural ending of a life and its transition to an afterlife. It is a difficult subject, one that we do not fully understand in humans. We feel ill at ease in advising what is appropriate for humans. In our kindred species, due to the sharing of thoughts and emotions, thoughts of extinguishing an individual's physical body is much more apparent and is less likely to occur because of the sense of connection that each of us and our kindred share. It would be tragic if, somehow, that safeguard was removed. Otherwise, in a terrible scenario, many of our species may wish to end their lives in a mass way. This would not be good if we telepathically shared the same impulse. This impulse is just what it is, an impulse. It should be one that is well considered and not spontaneous. We have seen in the human condition that there are times when humans will end their lives for reason that baffle us. We can read their thoughts. They are often confused and in turmoil that can be caused by a chemical imbalance. Perhaps there are hormonal imbalances in the brain creating a condition where humans wish to destroy themselves.

Are you speaking of depression?

Yes. This is what you call it. It is a hopelessness that is frightening to us. It is a terrible loneliness that we understand in humans, but our kin would find it more frightening to be disconnected from our brothers and sisters. We understand this happens with humans without a telepathic connection. What humans need to realize is that they are connected in other ways. Being telepathic is advantageous. It is not necessary for preventing such a depression, but it helps. We imagine ourselves in such a state, and it is difficult for this to happen biologically, but the despair causes us concern to see this in your Earth kin.

They can be helped if they do not take their lives. There are treatments for it.

I understand. If the victims take their own lives, it leaves no more options for the healers. It is not logical.

There are other cases where an individual may become completely incapacitated and does not have the joy of telepathy. In addition to this, they have uncontrollable pain. Sometimes these individuals choose with clarity, not to die, but to stop the pain through death. Their desire is to stop the pain and be free.

I have spoken of an irrational condition of the mind. This new situation you raise is different. It is one of clarity and rational thought. It is acceptance of one's reality where one may enjoy life no longer. In this situation, we can understand why one would wish to discontinue their life. We understand this and, reluctantly, we accept this.

I am sure if they had other options they would take them. No one wants to give up their life.

What you state is true.

Open Session for Comments or Questions

The lovely being (Zestra) that you brought with you this time, does she care to make any closing comments?

I sense from her that, even across different species, there is an understanding between females that transcends species.

You sensed that, and I did, too.

I sense this from her. I think you do, too.

Yes.

This pleases me. I do not pretend to fully understand, but I know you do. It is a lesson for the rest of us in finding connections between our kin. Zestra is happy, and I will leave that between the both of you. That is what I sense from her that she is sending you. I am here to respectfully serve and enjoy this.

We both have little smiles. Thank you for joining us.

Chapter Three

Session Thirty-Six
March 25, 2012

Steve is sent to continue his work with Scala, Lent, Ms. Peters, and others, on a planet-scaping project in another dimension.

Han, you are cordially invited to visit with us today.

Hello, Mary. This is Han. Zestra is also with me. Zestra is taking a quick tour of your house. She is noting the piano and the Monet painting, and your other things.

She must find it to be a very cluttered space.

No. She finds it filled with reflections of you.

Thank you. (Therapist reflects on this and reminds herself to clean the house before making contact with the Zetas.)

Earth History

We have just recently discovered that music, because it is a thought form, creates images with negative spaces, which surround the various tones. I can see that music would have a lot of different aspects to it, and we are just learning that now.

Music can evoke emotions and, in a way, create a kind of sculpture in the mind, a place, a feeling, or a thing, which is a thought.

Thank you. I couldn't find the right word, but you found them it me.

I have?

Yes. It is sculptural.

That is the thought that I felt from you.

Oh, thank you for getting the right word.

My pleasure. It helped me to understand your thought. It creates in the mind images and shapes. Music can often evoke or create beyond its original composition. When something extends beyond the original piece, new ideas and thoughts, and the creative process continues. Those who are inspired by their first work desire to create more.

Zeta Reticuli History

How does music from other cultures affect you emotionally?

In some ways, music can be a good barometer of each culture's degree of emotional and artistic expression. Some have more colors to paint with than others. Each strives for artistic expression. Each civilization is unique in this, but the range of emotions in song varies. The human range is quite extensive. In the modern music, of late, we have noticed that the emotional range of expression is more limited now, though it does not have to be. It is more a single beam with one specific emotion that needs to be expressed. Older music, of several hundred years ago, had a more dynamic range of emotions. It had many layers at the same time. It was much more complex, and that was more appealing to us. It had a range of emotions that we had forgotten, or we did not know existed. The music of Handel, Mozart, Beethoven, Tchaikovsky, Prokofiev, and other artists displayed a range that astonishes us and expands our boundaries. We find it is quite beautiful when we listen to it. Other music today is rather monotone in its emotional range. It is either one or two limited thoughts expressed.

That is true of the popular music. Some production of classical music is retained because movies that are made today usually require a full orchestra. That is, as far as I know, one major source for 'classical' music currently being created.

The classical music or similar strains are appealing and effective in expressing in very brief periods of time, the emotions required for a scene that can change quickly from scene to scene. The 'classical' music, as you refer to it, is more adept at fulfilling this in to help to express an emotion in the story, one of awe, suspense, excitement, despair. This form of expression gives us understanding about how you tell your stories about yourselves. That tells a lot about you and helps us to understand.

Sometimes the music is sufficiently composed that it can stand on its own, without the movie, just for its own beauty.

Music contributes to this collaborative art form. Some music, as you say, goes beyond and becomes a beautiful piece in its own right. As you say, it can stand on its own.

I could imagine that, between the various planets that you have explored, perhaps different species might come together and have collaborative efforts in music, art, and other kinds of cultural events. Have you observed this?

There are exchanges. There are cultural studies. One culture can influence another. At times, cultures are transported back to other worlds because they are aesthetically pleasing. Sometimes they are adapted. The Italian explorers brought back items from China. Other cultures have been transported around your world over time. It is not dissimilar from many cultures across space. There are worlds that have a distinctive style in architecture, and some seem modern or primitive.

Some species know how to express space and form in a very attractive manner, sometimes, in their complexity, and sometimes in their great simplicity. At times, to a complex mind, a simple clear concept can be very relaxing and easing to the mind. Art can be expressed in a simple, clear, clean thought, free of other extraneous clutter. For example, the perfect expression of a shape or a space can become quite admirable. These are often copied and taken to other worlds to be enjoyed. These cultural exchanges provide a basis for a greater understanding that help unite different species of sentient beings, kin, that exist in our galaxy. Art is one species' interpretation of a form in a space and its relationship in that space. How one being's mind expresses art tells much about the mind creating it, and how it sees itself in the Universe we all share.

Contact

During this last week, I noticed your presence very distinctly.

Say again please.

During this last week, I was aware of your presence at least once. I could feel it on my forehead.

You mean the energy in your hair again?

Yes. I do feel that, but there was also something a little different this time. When I listened to part of the tape from last week, each time I turned it on, I felt fluttering in my forehead, which is indicative to me of someone communicating. I usually contribute it to spiritual contact, but since it was tied to listening to the tape both times, I need to ask if that communication was coming from your area or perhaps from some other source?

You were able to repeat the experiment?

The second time I listened to the tape it occurred a second time. The fluttering is located in the area that we call the 'third eye.'

It repeated itself?

Yes.

Zestra is not aware. I am not aware.

Then it was probably spiritual.

It may be.

I am sure that others are interested, too.

Others are interested. It could be this. I do not sense anything to be of an ominous nature.

Oh, not at all.

It could be you have touched a spiritual side of us that maybe we do not fully recognize that you have sensed. You do have a great ability to detect things.

It came across as a curiosity. That was my sense about it.

I am now curious myself about what this was. As long as you are safe and protected, know that our species is protecting you, too. I do know that you are very capable of protecting yourself as well, and we would just join as allies, but I do not sense that this energy was anything but benevolent.

Well, I sometimes have a spiritual visitor called Djwhal Khul, a Tibetan ascended master stop by. He is an ancient one. He may have been a monk.

Monks can be very receptive to our thoughts. He may have heard our conversation and joined in from a distance, yet he was connected.

He listened for a fair amount of time and the second time was a shorter period of time. He is welcome. I can tell about his energy.

I am, for once, at a disadvantage. I feel amused. Sometimes I can sense things in you, but now, you can sense things that I do not see. I find that enjoyable and amusing, and it is very much a learning experience for me.

Well, perhaps we have our receptors tuned a little differently.

Yes. I find that interesting.

I enjoy sharing these things with you. Thank you.

I enjoy sharing with you. It is nice to learn these things, that there are other energies that pass through your home and through your mind and soul. Some species of animals have a greater sense of smell or hearing than we do. Perhaps your ability to receive certain spiritual energies is more acute to certain visitors than it is for most of us.

Thank you.

The Universe

You have heard of solar sails. I understand that there are spacecraft that use solar radiation for travel.

Yes.

I would enjoy hearing about that.

The solar sails are an efficient form of limited propulsion. It takes advantage of the natural solar winds that permeate from various stars. They can be used for interplanetary travel. They are not too practical for interstellar travel since great distances are involved, but this form of propulsion is more effective than your chemical-rocket propulsion. The more distance from the source of the energy, the energy collected diminishes, but the velocities that are acquired early create inertias that can be maintained for a very long time. Also, as craft approach other stars, their flight can be adjusted for braking. Delicate maneuvers are used for insertion into stable orbits. For interstellar travel, we prefer other means, but these are not foreign to us. Some space cultures readily use this form of propulsion, if one is not in too much of a hurry. In the greater scheme of commerce, some planetary beings regularly use this form of transportation, which for us is slow, but it is efficient in exploiting solar winds that many stars create.

I guess it would be like sailing on a lake, not for speed, but for the efficiency and the sheer joy of it.

Yes. It is an efficient use of the energy emitted from a star for the singular purpose of transportation.

Doesn't it get extremely hot in approaching and using these solar winds? I don't know how far out one should be to take advantage of these solar radiation pressures.

It is a very gentle pressure. It is one that is difficult to measure or experience, but a broad surface area can collect this energy and, given enough time, the craft can be accelerated to great velocities. Over great periods of time, combined with certain slingshot effects of various planetary trajectories, gravity can also be used to accelerate the craft when the two effects are combined together. Clever species use these forms of propulsion. It requires a great understanding of navigation, gravity, and mathematics. Maneuvering such a delicate craft requires a special understanding.

If you do not mind traveling relatively slowly, it is an effective way to travel. There are craft that use great sails to traverse. They slowly accelerate, using the solar radiation and gradually increase speed. When they come near the next star, they can again use the same pressure in reverse. Imagine something like great umbrellas or sails moving slowly across space. It is very beautiful.

You mentioned that sometimes you feel the music from other planets as a vibration, and it reminded me of the Earth kin. There are some who are completely deaf, yet they still feel the vibration of the music, even to the point that they can dance to the vibration.

Yes. This is true. We remember, in our study of Earth history, that Beethoven cut the legs out from under the piano and played the piano on the floor so the vibrations of the piano strings could resonate on the floor. This enabled Beethoven to express his genius, despite his handicap. The deaf can often feel these sounds. The loss of one sensory organ can be compensated for by other sensory organs, to a degree. Their skin becomes a form of sensory receptor of vibrations through the air. Some species live in the water, and some musical sounds travel through the water. For some species, it is through the air with each having its own tonal qualities, which can be beautiful. Some species use their entire bodies to both express and to receive the sounds emitted. Often, our telepathic abilities intermix with these sounds where we can almost feel the emotional intent of the sound being emitted. It is a very curious blend of sound and thought where the artist's emotions can be felt, as well as the sound. For some creatures that live on planets with oceans, these sounds can become quite sophisticated. If their visual acuity is less, they have enhanced their sound abilities and, as a result, many species will express sound better over great distances. Your whales have distinctive songs, for example.

Yes. We are quite fond of them.

They use sound for many purposes: to navigate, to communicate, and to stun their prey for food. These skills can also be used for music shared between them.

We find whale sounds pleasant to our ears.

They are mysterious sounds, like ghosts. They are eerie and evoking of curiosity, like from some great beyond in the darkness, carried by the depth and pressure of the ocean. There is life and wonderment from this.

Thank you.

I am glad we spoke of your whales.

I hope to see one someday.

When you see the eye of a whale up close, you will see (pause). Often our species stare with the eyes to see through into the mind and soul of beings. You will see this in the eye of a whale. You will see great wisdom, intelligence, and the bearing of a gentle giant. In order to know its thoughts and experiences, we have tried to communicate with whales, as we have tried to communicate with others of your species. There is another dimension that is vast and mysterious.

Isn't one of your joys in life to have some mysteries?

Sometimes it is not good to know everything. Sometimes, in order to relieve anxiety, mysteries give us new goals to accomplish with each new day. We do this in little steps. There is joy of having something new to do each day. Our mythologies in our culture are important, too, for inspiration and to complete our imaginations.

Oh, we are so fortunate to have imaginations.

We are not so different.

Open Session for Comments or Questions

(Han begins a conversation about wreckage left from UFO crashes and whether human scientists could back-engineer such debris. He expresses his view that humans could never properly back-engineer the alien technology they find.)

Our human scientists would not figure it out.

They have not figured it out, but they have kept it. They are waiting for a time when maybe their science will be available to fully exploit and understand it better. Right now, they do not have the capability. Also, if that time should arrive, it could easily be made to disappear by the aliens themselves, should they wish to do so.

For now, they have nothing to worry about?

Yes. We leave them with their toys. They still do not know where to put the batteries, you might say. A time may come that it serves a useful purpose because it challenges science to develop in ways that your science will one day match ours. Even though this may seem incredible, you can reach a level where we can communicate where our society and your society will form a balanced relationship, and your society is not forced to depend upon us. It would be able to exist in the community with other species. That day will come. It is beneficial that this wreckage exists. That is why it has not been already taken. It serves a purpose in challenging your science to understand it.

It is rather like holding the carrot over the nose of the horse.

Yes. The 'horse' is biting, and it will sample the carrot. One day it will benefit from the nutrients from that carrot, to use that analogy. It is a carrot. That is a good way to express this. I enjoy the way our conversation is heading. I know other human beings have many questions about this area. The crash wreckage does exist, and it is in various places. We know where it exists. We have maintained an inventory. Most of it has been recovered by off-world species who would consider such wreckage as too dangerous for you to have. We leave a few 'carrots' behind to tantalize your curiosity, based on a new science.

Man is not yet emotionally mature enough to handle this new science, but as you say, there are great strides being made.

Our technology is so far beyond human understanding that it is best that the current human frailties not grasp this technology at the present time. This is as it is.

Zestra, is there anything that you wish to bring to this discussion? I am about to bring this session to a close.

Han: Zestra is now communicating with you directly.

Zestra: I am Zestra, and it is lovely to meet you. I am enjoying the communication you have with Han.

It sounds like we share healing in common.

I am a healer at times. I am also a caretaker for our species and often the human hybrid species. I look after the humanoid entities of the human species. Sometimes I will communicate with human children and teach them about us. I plant thoughts into their minds for future times. They are ones of peace and harmony and mainly of understanding. We show them that we have a relationship. Children of human and alien mix become isolated from both. My role is to show that this not the case, and that they are very special. It is my work. I work with the Zeta kin children and the Human kin children when they are together. I am one who educates and teaches, as well as being a healer. Healing can be both psychological and physical. They are one in the same, particularly in our race. I sense the same with you. Maybe that is why I am attracted to your energy. You seem like-minded.

I think once you become a healer, you become like-minded with other healers. Also, I think we share that same belief that children are the ones that need assistance in terms of them understanding their world and universe. They are the future. If you want change, you start with educating children to be good universal citizens. I am pleased to meet you and to know that you are one who works in those areas.

Han's role is one of communication and diplomacy. My role is, as it is for many species, working with and caring for beings. Some of these you call hybrid or mixed species. They are hybrid kin.

'Blended' works for me now.

That is another term. I am enjoying multiple words for the same general meaning. It is like a music of words. I will step back and give Han your attention. I just step back respectfully and restore the attention to Han so he may address you and say farewell. My thanks to Han for permitting me contact with this human female. I understand. I spoke to you last time, saying that our femininity transcends our species' differences. We understand each other. With that touching thought, I step back.

Thank you. Han, are there any departing thoughts?

No. I wish you well until we meet again. I surveyed our journey today and the many topics we have discussed about both special individuals and in general. I wish to thank you again, and I look forward to our next meeting. I wish you Namaste, if we have your permission.

Always. Namaste.

Farewell, then, from Zestra and me.

(Steve is suffering some burning in his foot. Han sends thoughts and disrupts the pain signal.)

Session Thirty-Seven
April 1, 2012

Using hypnotic suggestion, Steve is sent to work with Scala, Lent, Nicole and others on a planet-scaping project in another dimension.

Are you available to speak with us, Han? I feel you coming in.

Yes. Mary, I am here. Nice to be with you again.

Thank you. You are most welcome. Are you alone today?

Zestra is not with me this time.

Thank you. I hope you spent your last week enjoying those things you love to do.

My colleagues, I am sure, will be visiting again from time-to-time in future sessions, Gen and Zestra, among others.

They are quite welcome.

They enjoyed meeting you.

Earth History

May I ask you about the Federal Reserve Bank? It is not Federal at all, but is privately held. I am wondering if that is an unnecessary institution.

I believe it is an institution for providing the financial integrity of many banks in the United States.

I am curious why it is private instead of Federal.

It is uncertain. I do not know why. It is a curiosity. It seems that there is corporate control of this in an indirect way. I am not familiar with the nuances of the American banking system.

I was just curious about it. It really isn't very important.

What was it that made you curious?

Because there seem to be so many secrets, so many hidden processes in the financial aspects of this country and of the world.

It does seem to be a secret society and a world unto itself.

Financial institutions have the primary goal to thrive, but even more than that, those who control them are not so interested in their customers thriving, but rather in themselves thriving. Anyway, it is just a concern, but I can let it go.

They hold great power around the world and within the United States. Such power is always a natural concern.

Yes.

Your curiosity is well founded.

I would like to ask you about another curiosity, a scientific one. I understand that the Earth's magnetic field has been growing for the last 400 years, but, now it is decreasing. Also, there is an anomaly in the southern Atlantic Ocean. We are not really sure why these events are occurring. Maybe they are cyclic, and our yardstick is insufficient. Do you have any thoughts about this?

The electromagnetic fields of the Earth do fluctuate on a cycle. The sciences of your world are beginning to discover this. In some ways, it is affected by the moon body, which orbits your world. It is a moderator for many things. It adds stability to your planet in its rotation, its angle in relationship to your sun, and, as a result, maintains a steady, consistent climate for life to develop in a good and efficient way. With the magnetic fields, there is a relationship between these forms and the Earth. There are many effects, but the fluctuating magnetic field is a normal pattern.

Thank you. It is our lack of perspective over time that limits our understanding of these events.

In time, as you record more data, it will be one more interesting phenomenon to study about your world, which can also be applied to other worlds in your explorations.

You know about the many mammals that we have on Earth, including elephants.

Yes. The 'great towering ones,' from our perspective, as well as yours.

They have amazing ears.

This is dependent upon the species of elephants. Some are smaller.

Oh, yes.

African elephants differ from the Asian elephants.

I was curious about their attachment to the bones of their deceased kin. Do you know whether or not elephants are aware of death? When they smell and caress these bones, I am making the assumption that it is like coming across a sweater or something that a loved one used to wear and the scent bringing back memories of the deceased. Do you have any insight into elephants and their concepts about death?

Yes. Elephants have a very close-knit social order in their herds. They are very protective and have many close relationships within the herd. There are leaders and followers in a number of family groups. When a loved one grows old and dies, or an elephant is killed, it is a great loss to the herd. One of their means of identification of one another is imprints of memories, among many things. With their sense of smell, they can have very vivid memories, as humans often do, as you have reflected in your question. When one of their herd is lost, it is felt among all of its members. We can feel this energy among them. There is a very close relationship between them, and the death experience is felt more by them among their fellow kin than among other species, to a degree. When poachers are exploiting the herd for their tusks, it reminds us of the entities that seem to harvest the human genetic information with disregard for life. It is one that is of concern to us. They are also poachers.

Most of us are very disturbed about this poaching.

The elephant is a magnificent animal and has a highly developed devotional order. They hold great majesty over their domain. Their herds are small, and they struggle to survive. We must, at the same time, detach ourselves from what often occurs on your planet. Some of it is distressing. Some of it is encouraging. There is a certain detachment that we must retain, yet we express the need to impress upon humanity, overall, the continuing and necessary duty of your species to be good stewards of your planet. You do not have entire control of your world. The forces of the environment of the planet itself are, to a large extent, maintaining certain order. Human influence can be invasive and can often displace other species. The very shallow need by some cultures for elephant tusks is so unnecessary, at the expense of the magnificent elephant species on your world. We have also transplanted species of elephants on other planets as well. You might be surprised.

I hope they are thriving.

They are kept on suitable planets where they are comfortable and thriving, and perhaps the populations can be rejuvenated on the Earth planet.

Since the thing that makes them attractive to a handful of people is their tusks, and I am wondering if nature will change the genetic code in the elephant so that those tusks no longer grow long, or are so insignificant that they would no longer have any value to humans.

That is one possibility. The change of genetics over time may be too long because the species could be wiped out in the meantime. I mentioned the importance for our species to be detached while observing all of this. It would be an inappropriate use of passion to destroy those who kill these animals. We must step back. A violent intervention would be a violation of our principles, but the motivation is there. This would be an inappropriate use of force.

I know you must go through this in your many interactions with many worlds and species. I could see how it would be overwhelming if you became emotionally involved.

Perhaps this is the reason we have survived this long. We have detached ourselves from our passions and emotions. On an interpersonal level, we have grown accustomed to this, but when we are in contact with other species where there is contrast, we may seem cold and analytical to other species. However, it has been one way that we have learned to survive.

It is very understandable.

When we come in contact with humans, and we remember our ancient ties, we become aware that our emotional range can seem limited when compared to humans. It is perhaps best, in the long term, that it is this way for us. It has been successful for millions of your years.

There has been some unusual activity in our national parks. There was a discussion on the radio last week about the National parks and the missing persons. This includes Canada and the United States. It was stated that there are clusters of disappearances. One of these areas was in the Philadelphia area. Some of these incidents include dogs tracking the scent of a missing individual and then suddenly stopping and refusing to go any further. Other cases involve children being found many miles beyond any place they could go by themselves, i.e., across raging rivers. Some are found with their clothing turned inside out. When one tries to find out how many people are missing from a certain park, the park authorities claim that they keep no records of missing persons. There have been disappearances at Crater Lake, Oregon. There are lots of people missing at Tenaya Lake in Yosemite. Because of the peculiarities of some of them, clothes turned inside out, etc., I believe some of them are abductions. I know that human beings are careless. Are national parks a favorite places for abductions, or are we mostly seeing careless humans?

Many humans, as you mentioned, just disappear, and the ability of some humans to find them is limited in the vastness of such areas. At the same time, the conditions pose an opportunity for some illegal species to abduct on one-way trips for their own selfish purposes. These are often, as you refer to them, 'good fishing holes' for kidnapping and making humans disappear. There are also some portals around the world where time and space become doorways to other times and space. A number of these exist all over in Michigan, Pennsylvania, California, Utah, and other places. The Skin Walker Ranch is one example where there are doorways. There are energies and entities that often use these portals. They can be both benevolent and malevolent.

The malevolent ones are to be watched. They can often take advantage of individual persons who wish to get away from it all by going into such wilderness regions. They often suffer the consequences. This is the reason. Sometimes there is a jump in time and space where it would appear that a child would be unable to ford a dangerous river. In some instances, they made that jump in time and space by entering a local portal that enabled them to be on another side of a river. Or, the entities, using that portal, have taken these people and transported them for whatever purpose.

I knew some were gone forever, and some were returned. Of course, wilderness can be a dangerous place for most people.

As humans came to live in social communities, they began to lose their instinctive abilities to live in such remote environments. They become incapable of surviving due to their dependence on the infrastructure of their societies, including their built-in notions of safety. I had to carefully choose my words to express this. It is all a matter of perspective. I think you understand.

I do. I was aware that there were portals because I read stories of people crossing fields and disappearing right in front of others. The voice of the person who disappeared could be heard for a little while and then silence.

These are small crevices in the fabric of other dimensions with which we are familiar. The quantum physics is similar to what we discussed in celestial navigation and transportation. One is like a cave, a natural formation in the structure of the Universe versus my earlier analogy of a deliberate tunnel through time.

I was curious about the behavior of a cell in a Petri dish. Apparently, if one places a single cell in a Petri dish in solution, it will immediately start sending out nano tubes, apparently seeking another cell. Is it at this level that the desire to connect with other life forms begins?

The desire is different than it is with life forms that herd together for safety, although this is part of it. There is a desire to connect at a bio-chemical level to provide a mutual form of strength in shared chemistry. It is at that level, not so much an emotional connection or a desire for belonging as it is simple chemistry. Yet, as we know, we are all chemical beings and, at some level, these processes can seem instinctual. There is a desire to join with others for mutual survival. It is a curious thing to watch.

Is Fukushima creating any significant nuclear contamination here on our west coast?

You refer to the damaged nuclear reactor. Yes, contamination is found, but the news is very suppressed about it.

The Universe

(Time is discussed with the Zetas and how they observe time.) The therapist asks Han, Are your years different from ours?

Yes. The time is the same, but it is measured differently. You use a year as one orbit of your sun. We have a longer time cycle of about 2.3 of your years that it takes us to orbit our primary sun. There is also a secondary sun that I have talked about: Zeta Reticuli 1 and Zeta Reticuli II. Our, as you call it, super Earth, orbits around Zeta Reticuli 1 and has an orbital cycle of 2.3 years to your 1 year of orbit.

Orbits would be a common way to measure time across the Universe?

It is in the beginning. With the introduction of interconnections between other planets, space travel, and using quantum physics to leap from one area of space to a vastly different other space, we distort space itself through quantum means. This has required, literally, a Universal time, which is separate now from many local planet-time measurements. This Universal time is used to not only pinpoint navigation through space, but also pinpoint navigation through time. It is necessary to facilitate commerce. Much like your Greenwich Zulu time as one time measurement, regardless of where you are on your planet, we have an equivalent time that is used by most species who have agreed to this concept. This Universal time concept connects the many thousands of planets in our community with many species so that there is uniformity in time reference. It is used for navigation between stars and planets.

It makes perfect sense. What about interacting with other dimensions in this Universe? How does Universal time work in these other dimensions?

We use other dimensions as the links to other distant locations. Intermediate jumps are the means for this navigation. We exist in the same dimension, but we use other dimensions to travel in astonishingly short times.

Are there other dimensions within our Universe?

Yes.

And there are other Universes with their own dimensions? How do you manage to come up with a commonality of time in other Universes with their dimensions? Maybe it is not necessary.

There is always a need for a time reference. There is a beginning and an end point in any journey. There is a measurement of time that is universal within the very smallest elements of space itself that is common everywhere. This acts as a universal clock for measuring time and gives a reference, regardless of time displacement. Perhaps it is a difficult concept to explain, but it is necessary to know where one is going and where one has been and to travel on regularly scheduled journeys between worlds, both for communication and commerce. There are great trade routes that require such measurements. We do not often think about this, much the way you do not think about the clock on your wall, but when considered, it is a very important aspect of our civilization as a spacefaring species among other species. Each shares this form of time measurement.

Is time also tied to space so that it is a time-space continuum?

Yes. The two are inexorably linked.

Is this time-space continuum contained in the very smallest elements of the Universe?

Yes. In all elements. It exists everywhere and provides the basis for commonality, which can provide a standard time measurement, like an atomic clock.

It sounds like a multi-universal baseline.

Yes. It is much faster and more complex than your concept of time. We leap between dimensions where a dimension is what connects the beginning and the destination. In between, it provides another dimension, a system of universal time, even within this other dimension. It is necessary to be established so that when the dimension used is quantumly distorted to provide transport, we emerge from it at our destination and the local destination time is accounted for in the flight.

We discussed the psychological effects of moving through this kind of a process. This additional information is helpful in forming a larger picture.

In our past discussions, we have talked about psychological effects of spaceflight, which is also very much time travel, because time is distorted as a side effect of such travel. There is also a technological concern. It is all part of the fabric or the navigational infrastructure to facilitate such travel in a way that can be practical for specific species and many other species at the same time.

It is rather curious to me that, when one dreams, the time-space reality seems to fall away. It seems to be irrelevant. Do you experience that also?

Yes. Our concept of time changes. One literally lives for the instant, and the concept of aging can seem rather abstract at that point.

It is curious. I don't know why time-space is unbound from the dream state. Perhaps it liberates us because we need to be temporarily unbound from it to truly explore.

Time travel is much like a dream. Both can be liberating. One is free for a time from aging. We live for the 'now,' in the instant. Abductees Betty and Barney Hill puzzled the abductors when Betty commented that her husband's dentures were what happens when humans get old. This was a very interesting concept to that species, to which we are related. They did not understand what 'old' meant. It was a very strange term. They needed to remind themselves, being space travelers who also time travel, that there is an effect of time by those who are relatively stationary in their position in time and space. When one describes those who travel beyond the speed-of-light, they do not actual physically move beyond the speed of light. The quantum speed of light can be used as a measurement. It would be many times faster than the measurement of light to measure the distances traveled in such a relatively short time. However, we do not go faster. We distort space to make the distances shorter.

What term do you use instead of the light-years measurement to convey incredible distances?

I am searching for the word in our language. It is related to your 'light year' or velocity of light. Light did, at one time, early in the

Universe, travel faster than it does presently, but it is still a useful measurement. You use the distance light travels in a year. We use a measurement of dark matter. The velocity it takes to travel a distance between a known distance within our measurement of a year. The velocity of the physical Universe can be used as a measurement gauge.

Thank you. I appreciate learning that and always appreciate learning about the Universe.

We use similar concepts, but different measurements.

I thought they must be different for you.

It was once the same as it is presently for you, but space and interdimensional travel required a whole new form of measurement. The challenge was not to establish the measurement or even to technologically achieve that form of space travel. The ultimate challenge was a political one in getting many civilizations to agree to one standard. As usual, we have our own problems, but this has now long been accepted. Now it is the standard. This is what is.

When Europe was attempting to agree to a single currency type, the Euro, I was astounded that it was accepted by most European nations. I am trying to imagine the difficulty that was faced in getting cooperation from thousands of worlds to accept a single new form of measurement.

That is a good analogy. In both cases, ours and yours, related to the Euro, mutual need to agree was the driving force. A practical solution was necessary, one of mutual benefit. That is another example of an analogy that illuminates. This is a very interesting topic, celestial mechanics and navigation, that we have touched upon.

It is certainly interesting to me because it is all new to me. I have a deep curiosity about the cosmos, and now navigation in the cosmos. I have fondness for these things, but not much knowledge.

It will extend. It is the same with me. We have our navigators. Yet, in a way, our craft are also intelligent entities. These craft are existing conscious beings with enormous technological capabilities. Much of the necessary, mission-critical functions of our craft are built into that intelligence. It is akin to you dialing a phone number. You know the phone number, but you do not necessarily need to know the great infrastructure that is required to connect a phone call through many complicated mechanisms. In a way, this is the same for our stellar navigation. We merely think the thought to our spacecraft, and the craft makes the necessary arrangements to travel to that destination. It is not so dissimilar. It would be easy to take the great complexities for granted. It frees our minds for other things.

In your relationship with these craft, do you feel, not so much a kinship, but perhaps a sense of affection?

They are technological living forms that are self-maintaining. We must sometimes provide additions for maintaining these craft. There is a psychotronic connection, as we have with many forms of technology in our society. The craft become like an extension of our bodies, in a way. The relationship, as an extension of ourselves, might be a way of describing how we regard our craft. They are not living, organic beings, though they do have certain organic filaments to be receptive to our thoughts. They do mimic our minds and respond to our thoughts. They have an artificial intelligence. There is a form of protocol for survival and self-preservation. What is good for us is usually good for the craft. In many cases in our travels, we serve our vessels, and our vessels serve us. There is a mutual connection. The vessel also acts as an extension of our own senses. It makes us aware of what is around us in such a way that a human pilot would be very envious of these capabilities. This facilitates us to be able to see and sense farther. It allows us to project our thoughts toward humans as we hover above the surface of the Earth from a great distance. UFOs may sense a desired contact from humans. On occasion, when humans have correctly distinguished one of our craft from a natural object, they will send light signals. Sometimes our craft will respond. We respond to study and to entertain. We will measure the minds of the humans we are interacting with to sense their disposition toward us.

Is it the craft that is actually sending the signal in response to human's signaling with flashlights?

It is both our occupants and the craft in unison, working together to make the lights go on to respond to the human stimulus.

It doesn't sound like there is any need for an emotional attachment to your craft by you.

We have an affection for our ships, much as you do, but it is a limited emotional one. Call it pride in our vessel. The craft's intelligence appreciates this. It can sense this. There is, in a sense, a relationship, a very limited one.

Do you find it curious than male pilots and captains often assign a female gender to their planes and ships?

It is to express an identity or personality to a technological device that often has certain performance characteristics. However, in some of your cultures, we have noticed that some are also addressed in the masculine, particularly in very large ships where it is felt, perhaps, by the male of your species, inappropriately, that such a large vessel can only be addressed as the male. Germany and France have such concepts for their very large ships, for example. It is just a curiosity of those cultures. Even their languages speak of objects as masculine or feminine. It is the way that your species relates to many objects in your reality paradigm.

We do not refer to our craft as a male or female. The craft is an entity, but also an instrument that serves us by facilitating travel and local study. Often these craft would easily travel without occupants to perform their many missions. This is often so.

It sounds like it would take a tremendous effort to build one of these craft.

Many of these craft have the technology and the intelligence to create their own forms of craft. Our economic system is one of exchange or barter in such enormous volumes of energy and resources that would be difficult for Earth beings to comprehend.

You trade with many planets?

Your civilized development is only at the beginning of trade between the nations of your planet. We, however, are on a different level of magnitude where there is trade between many other planets beyond ours. In the myriad stars, there is material available for the construction of craft and supplies to harness great energies that are necessary to travel. The energy is relatively inexpensive, and travel is relatively free, but it is all relative.

Are you familiar with a deeper level of gravity that has to do with telepathy and a kind of 'half state?'

There are forms of 'super gravity' that can distort space and time. Much telepathy is a quantum notion. It can also be distorted by such super gravity anomalies. We have some in the form of what you call black holes. But there are other areas of space where space can seem like a lumpy porridge. The density and gravity spectrum can be seemingly different and 'lumpy.' Sometimes, across distances, telepathy, which requires a form of quantum physics, can be affected by these anomalies.

Is it a temporary effect or a long-term effect?

The effects are temporary locally, but long-term everywhere.

Other

Is there anything that I can do to more effectively facilitate our communication?

Steve is somewhere else, but your voice comes into his ear, into his mind, and then I 'hear' his thoughts. And I 'hear' your thoughts, also. However, this is 'hesitated.'

Yes. I know that you can come directly to either one of us. I am looking for methods you might have that would make it easier for the communication process. I am fine with the process. It comes in clearly, and I can easily transcribe it, but I also don't mind if it is a little faster or perhaps sometimes multi-layered. How can we make it better?

To improve our method of communication?

Yes.

If there was direct link between us, I am just wondering, with your technology, how you would be able to record this differently from what you have done in the past.

A direct link with me would not work with our technology. The improvement in speed and layering would need to be done through Steve. It is not necessary, however. It was simply my desire to make the communication flow easier through Steve.

This will be considered. Multi-layering can occur through Steve, but the transcription of it would be a single layer.

You may be aware that sometimes I take clients under hypnosis into the future. How far in time could I comfortably take them so that they may get useful results? A year or two? I know that you can see into the future to some extent. I usually keep them to two years or less.

That is about an appropriate time. Anything beyond this can become impractical to define the knowledge that is gleaned from this process when they return to their current reality.

I did it instinctually, without really knowing at what point it would be less effective. I didn't want them journeying too far or observing things that they did not need to observe.

There is too much randomness. It creates exponentially alternative pathways. There are so many routes in anyone's journey. It is best to just look at the immediate choices one may have, though a limited time frame of journeying is appropriate.

Humans do not do all that well with the concept of probable realities, so thank you for that.

Open Session for Comments or Questions

I am going to be working with Steve to enhance his ability to do remote viewing. Do you have any suggestions?

Interesting. I encourage this.

I know he can do it.

Yes. He can. He just needs some initial assistance.

Is there anything that you would like to add to our session? By the way, I enjoyed it very much.

I enjoyed talking about the space travel. It is one interest that is close to me. I enjoyed it very much. Beyond simple travel, to be in space is special. This is a universal concept among our species. I have enjoyed our session. We have talked about some fresh subjects. In a way, our discussions have provided further insights about our spaceflight infrastructure, our communication, and navigation in spaceflight. Due to the means of our technology, our craft have their own artificial intelligence. There is a harmony. Only the new trainees argue with our spacecraft.

(Therapist laughs.)

I am looking back on my early days and our other colonies. It is an amusing sight. I have no further comments at this time.

It is the way of the young to question.

Yes, it is. Due to self-preservation, these spacecraft also question the young so that the craft may continue to preserve themselves. Such are the young everywhere, universally. Fortunately, our craft can protect themselves from new and inexperienced travelers.

Thank you for that lovely sight.

It is an amusing place to end. You have your high insurance premiums for your young. We have our spacecraft, which argue with the inexperienced and will supersede the commands. Our craft can easily drive themselves. As any young creature in developing relationships, either with their species or a technological 'species,' there is always the intellectual and emotional development that must occur before attaining maturity. I enjoyed having provided many new insights, and at this point, I am concluded with our discussion, unless you have other things you wish to discuss. I am at your service.

In our next session, I would like to ask about what exists beyond the spiritual.

This is a tantalizing question.

With that I bid you a fond adieu. Namaste, and my greetings to your friends, too.

Namaste to you. We wish you a good week.

Thank you.

Session Thirty-Eight
April 8, 2012

Steve is sent to work with Scala, Lent and others, on a planet-scaping project in another dimension. (This is a mechanism for removing Steve's awareness of the exchange between Han and the therapist.)

Han is cordially invited to speak with us today.

Hi, Mary. How are you?

Thank you. I am quite well. It is incredibly beautiful here today.

It is also the same here. I am sitting here with Zestra and Gen. They have come here to be with you on this day.

That is wonderful because I have all kinds of different questions for all of you. Of course, I also enjoy just one-to-one interaction.

We interact freely today. It is the three of us, and each of us can speak. This is Zestra. This is Gen. The three of us are speaking freely, so I hope this communication works well for you.

As each one chimes in, we will all enjoy the process.

I will identify each of us so that you can understand the difference.

Thank you.

Steve has also been given instructions to relate on many levels, and that is more comfortable for us. It may seem strange to Steve, but I think he will be able to adjusted.

Thank you. I see it as akin to playing piano music where there are multiple tones played with two hands, tones all streaming in together.

This would be more comfortable for our thought processes. Thank you for this consideration.

Yes. I was hoping to bring a little more comfort for you in the communication process.

Earth History

Can you tell me what has been happening with the dolphins here? We have had large schools of dolphins die off suddenly. I am wondering if it is tied to sonic or acoustical testing by the oil companies who are testing for oil.

(Long pause, apparently for a discussion between Han, Zestra and Gen.) There are often tests in the ground under the oceans that are done acoustically to find these oil deposits. These tests often reverberate into the oceans. Also, there are ultra-low frequency sounds and tests done by naval vessels, which are conducted for narrow purposes for the protection and navigation of similar ships under water. They sometimes confuse and confound the species known as dolphins, porpoises, and whales. They use and depend on acoustic sounds for their communication and navigation and added abilities, which sometimes have been thought of as telepathic between their species. They have a sophisticated, telepathic-like form of communication that exists in their species on Earth. The sounds can become confusing and disruptive. They often lose their ability to navigate and find themselves stranded. This has been a problem that has affected them, and they have found themselves in trouble, floundering on beaches. This has been distressing.

So that is the unknown cause of beaching.

Beaching? Thank you. I have been trying to describe in a long sentence what you say in one world. Thank you for this efficiency.

Sometimes we try to call the whales back out to sea by using recordings of their sounds.

You try to lure them back into safety.

Yes.

We appreciate this. We see this among the humankind, and we understand and appreciate these attempts.

The whales also seem to understand because we will go up to them in groups and try to push them back into the water or pour water over them. They don't fight us. They watch us, and I guess they are sensing our intentions.

They sense, understand, and appreciate this. The pouring of water on them is an expression of not letting their skin dry out in the sun. It is soothing, which they also understand as a kind of stroking. There is a beautiful, sensual stroking from pouring water in your desire to return them to where they feel most secure and comfortable in an environment where they belong.

Years ago, a young whale came up the Sacramento River and became lost and confused. We tried with boats and noise to get the whale to move back down the river, but he just fought with the intrusion of the boats. He was here for weeks. We thought he would die, but so many of us were praying for this whale. I don't know how or what happened, but that whale finally did turn around and worked its way back to the sea. We rejoiced. The humans called this whale 'Humphrey.'

Humphrey?

Yes. It was Humphrey, a young whale, which had its sense of pride in thinking it knew better. It is much like a human feels when another human thinks he knows how to drive better. All were working for the same goal. Finally, the young whale understood and submitted to the navigation of its hosts, and it returned to the sea.

It was in fresh water, silted water. The water was silted. It affected the buoyancy of the whale. It was strange. Its buoyancy was affected, and it felt odd, since fresh water was not its normal habitat. It finally turned around in the din of the noises that were confusing it, and it began to understand. It became tired and relented, due to the sounds. It gave up and obeyed the sounds and followed them. Eventually, it did find its own way.

Well, that whale will have stories to tell its children. (Therapist laughs.)

We are all so…. sounds of other…. I think even saying…. (This confusing sentence reflects all three Zetas talking at once.) The young whale thought a submarine was its kin and sang to it. When it did not respond, it still saw the kinship in it. It sang to the submarine nearby that was heading out to sea and accepted it as kin. The young whale continued out to sea and then eventually found its own pod, and it was happy.

How did it know the submarine was going out to sea?

It observed where it was going. It sang to it, but it did not reply. It was not one of its own, but it was similar to it. It was a familiar shape, and it was also heading in the same direction. It was a human allegory of its own kin. The submarine did not make the same sounds, but the child whale sang to it. It saw its purpose to be a follower so it followed, to a degree, and eventually found its way out into the ocean and appreciated it. In the deeper water, it reacquired the salt water and eventually found its family in the ocean. Humphrey was back with his family.

Quite an adventure.

He had stories to tell, and he was missed. It felt good to be home.

Thank you for those lovely insights.

As you know, Yellowstone National Park has been having swarms of Earthquakes for some time now.

Yes.

Has that been going on for hundreds of years?

For a long time, and increasing.

I know there is a little ground swelling, and eventually it will erupt.

There will be an eruption.

Yes, as Earth is prone to do. Reno, Nevada has had swarms of Earthquakes, but that area is less well known geologically. I don't know if this is the result of magma under the ground or if there are some faults we didn't know about.

Around Reno?

Yes. Reno, Nevada.

It might be related to the natural folding of your plates, folding inside of each other where one goes lower and the other plate slides over the top.

We call that process 'subduction.'

Yes. There is that. I see in my thought. I am sorry. I did not state that well. There are layers of thoughts that might seem confusing. There is a rolling down and a rolling over the top in that location, in that general area.

The subduction goes further into the continent than I had suspected.

There would be subduction in that Reno location.

Thank you. I will keep that in mind when the swarms come again, as I am sure they will.

Yellowstone, too. Yellowstone is a remarkable place.

I hope I get to see it someday.

You will.

May I ask you about a mythological figure called Moses? I don't know if he was real or not.

He was a real person.

He was apparently sent to a mountain top to talk to God and, after some very strange interactions, came down from the mountain with stone tablets containing what we call the Ten Commandments. I am not sure how accurate that story is either, but I am wondering if, perhaps, he interacted with an off-planet species when he went up the mountain.

(Long pause.) The Ten Commandments are expressed values that can be shared in common with many species. There are certain fundamental precepts that are shared with us. If this happened exactly the way it has been described, I refuse to state anything different, hoping to respect the beliefs that are shared locally, but they are ones that are, in a broad way, not different from ours. Forgive my not being specific.

His interactions on the mountain were with a light/heat form, according to the story, and it may have caused radiation burns.

His white hair?

Yes. It turned his hair white.

The streaks?

Yes. That is why I wondered who he was actually meeting with. These commandments were apparently carved into stone, which probably would have required something like a laser tool, due to the time, unless Moses was very adept.

Well, I am deliberately being non-specific, but I do know that I do not wish to share beyond this. I do know that we embrace many similar values. *(This is a very telling statement.)*

Most of them are what I call functional values that enable a group to get along together.

Yes. That is the way we share them, along with you.

I remember watching a Mel Brooks' comedy about Moses. He comes out of a passage in the mountain, and he is holding three tablets and starts to talk about the fifteen commandments.

(Han laughs.) I thought the of the same joke.

Yes. It was funny.

I know. I thought of sharing that with you.

(Therapist laughs.)

I did not know how far I should go with you.

Oh, it is okay. I have a sense of humor.

He dropped one of the tablets and then presents the ten commandments instead of the fifteen commandments.

(Therapist laughs.)

I know the one.

It tickles my funny bone, which is a curious expression.

I know the one. We share the same joke.

Zeta Reticuli History

May we talk a little bit about your home planet?

We would like to.

Would you care to describe some of the life-defining geological events on your planet? I am talking about the times when you were already sentient beings and the various geological forces that forced you to make certain decisions that would affect the way you lived.

The way we are?

Not so much the way you are now, but the way you were when you were in the tribal state and maybe a few hundred years beyond that. For instance, here on planet Earth there have been massive Earthquakes and tidal waves that affected humanity.

Our origins came from birds, you remember. It gives us our physiologically, our bone structure. (Long pauses for discussion.) We eventually formed as land creatures and birds of the air. Eventually, we became more land creatures than creatures of the sky. We were able to move freely around the planet. I am trying to search for the word....migrate. That is the word. We could migrate. Some settled in certain areas with others, yet they could move freely.

Could you predict the geological events and avoid them?

Yes. In a way. Much of our migrating was tied to our planet to facilitate navigation. When certain geologic occurrences happened, this affected our navigation. Zestra is speaking. We knew how to adjust to these changes and navigate/migrate. There was a usual migration pattern. We were predominantly free over the lands, but we could also travel over the seas. In time, we became less aerial and more localized. As we developed and evolved, this ability to migrate, with a sense of being in touch with the planet, helped to develop our telepathic abilities.

Including a telepathic connection with the planet?

It included a communication with the planet. Because we were able to sense its magnetic fields and abilities in navigation, it helped us to develop our telepathic abilities with each other. It slowly evolved and developed at such a rate that it was almost imperceptible when we began realized how much we were in communication with each other as much as we were with the planet. This is strange. It just kind of happened naturally. It kind of grew upon us before we knew it.

Then it was a symbiotic relationship with your planet.

Yes.

That would explain why you were able to avoid what I call natural disasters.

Yes. This is what happens in a symbiotic relationship. As such, it was a profound time for us. We had different societies, but we eventually grew into one as our telepathic abilities grew. We formed greater links, and it was a uniting force where we were really one with each other in a way that was quite profound. This was a time of greater purpose.

Did you also communicate with other creature forms around you?

Yes, if they had the same type of unity. There was a closeness and appreciation, primarily with our own, but there was a sense of kinship with the others.

There were probably groups, or herds or pods that worked in unison in those days.

We could see this, to a degree. There was a greater strength of connection with our own, but there was also an appreciation and connection with other species.

I noted that you came from birds, and that the joy of flight has never left you.

It has been a deep, instinctual feeling that dominates a lot of what we do and our desire to travel (and migrate).

May I ask about a covenant among the Zeta kin and also between and among the other non-Zeta kin where you have regular commerce across the many inhabited planets? I know there must be some agreed upon types of behavior, a covenant to facilitate commerce. Being telepathic, so many of these requirements for behavior are obvious, honesty, for instance. Could you clarify a little bit what the behavioral expectations are before a species is permitted to join in commerce with you?

There are certain behavioral covenants that exist in this arena that we share with many species. Some of these behavioral and social values include respect for life; a desire to survive, but more a covenant of mutual understanding; a covenant of shared survival; and the honesty that naturally comes in communication. There is a desire to live. There is a desire to be happy. Common attributes of many species include appreciation, beauty, love, caring, curiosity, and compassion. In having these common goals in their societies, it is easy to create a form of mutual understanding among many species.

Do you meet together as a kind of commerce federation to facilitate this?

There is a prime group or confederation that exists that is among us all, and we are all a part of it. We are part of this confederation. Many species and kin belong to this. It is a kind of a group of many worlds that share in this. We have talked about other species sometimes and how we have been unhappy with species that exploit certain other species, such as a species that abducts humans. That is just one issue that we have discussed. Other species have shown the same unhappiness in humans being abducted illegally, and we often share this in common. There are many species that do not like it, so you can take heart in that.

Han clarifies the setting: Throughout this whole conversation, I, Han, have been in the background. It has been mostly Zestra speaking, but she speaks for all.

Thank you, Zestra.

Gen adds his comments: Han and Zestra each share the same commonality, but it is just for your pleasure to know that mainly Zestra has been here speaking today. We all share in this, Han, Zestra, and myself, Gen. We belong to the same community of mind and share that commonality for these sessions.

It is just pleasing to have all of your voices.

Yes. We all belong to the confederation.

Spirituality

I think this might be a good moment to interject the closing thought from our last session. It would be enlightening to have various perspectives. We had talked about what might be beyond spirit, and I gave it some thought. Also, I had to ask myself, 'What is the purpose of spirit?' I contemplate what might be beyond it. You have already met a species that is pure energy and, perhaps, a kind of spirit. It is one that you respect. What kind of thoughts come to mind when I ask: What is the function of spirit, and what might lie beyond it?

Oh, this is a beautiful opportunity. In the past we have spoken only through Han. (Zestra speaking.) Han has spoken for Zestra and Gen, but he wishes to relinquish that. There is instead a free exchange, with each expressing his or her own thoughts. I am so glad this has been facilitated because of the thoughts and ideas that are being exchanged here, we feel freer in each sharing our own thoughts. There is a beauty to this and particularly with these ideas and notions. In discussing other species, there is this particular kind, which is more energy than physical in concept. Still, energy is the same as physical. Your Einstein knew this. Matter is the same as energy. Energy is matter. There is an equality in this in their values, and yet they are different, and yet the same. The beauty is the same. They are not necessarily separate, although they can have properties that may be separate. They are still equated the same in their relative values of amounts or proportions. We are all thinking about this (small chuckle) each in our own way. (Amused.) It is funny how we are each greater than the sum of us, and I think that is part of the characteristic of our species. It creates a whole that is greater than our parts.

We have often pondered this 'pure energy' species, and it is with awe that we try to comprehend it. It is, in a way, a spiritual meaning, where the physical ends and the spiritual begins. In the transition of life, and at the beginning, when spirit is introduced into our physical reality, there is this mysterious part, which we have deeper glimpses into with our telepathic thoughts. We might have a deeper understanding by being able to witness it. Through our thoughts, we can share these experiences. We have deeper notions of where one comes from and where one goes. The ultimate point of origin or destination is still a mystery in these transitions. With the species of pure energy, we find this serious and tantalizing in its essence. We wish to understand more. We all find this disappointing and frustrating, yet the mystery draws us. We wish to know more, and we are all nodding in agreement with this. What do you think, Mary?

I have some thoughts and some experiences.

We would love to know them, if you are able to share these with us. *(They already know them and are just being courteous.)*

I am certainly happy to do so. I think, perhaps, that we can all agree that evolution probably plays a role in spirit. By that I mean living out life in a physical form, retaining the memories, and working from that base of memories for the next physical life, if one chooses. Having mediumistic abilities enables me to have some understanding of the passages in birth and death. There are other experiences in between. I have experienced souls coming to me many times. Sometimes a soul will literally pass through me and, at that point, all of my senses are heightened. I recognize the energy, as you described it before. It is like pouring water out of a vase and all of a sudden it becomes like an ocean. One becomes aware that the soul is massive. Also, unless there has been some distracting illness, the soul is compassionate. There is something almost musical about souls in terms of giving the vibration of incredible harmony.

We are all agreeing with you. There is this exchange. We also marvel. There is an irony for us that we often do not express emotion. I have been noticing how the three of us have been sharing feelings more with you, while you have been sharing more of an intellectual form of the same thing with us.

I think the soul has somewhat different or additional sensory apparatus. For me, it comes through as though it is playing on the 'strings' of my emotions. My intellect is also there, but the beauty of it only seems to come through my emotions. Perhaps that is the little variation I am sensing. The beauty just takes my breath away.

It is like we understand among our species, too. Between humans, almost like lovers, souls exchange and share something. You have done this, but you have given it another dimension. Using an intellectual view, whatever its proportion, we are impressed. It gives this whole passage of souls a deeper understanding. It is one that we are trying to understand in this species of pure energy.

Do all of you at least have some consensus about it involving an evolutionary process or do you think that may not apply?

It is so, but it is multi-dimensional. There is a linear-time dimension with passages at the beginning and ending of the physical life, while it also exists in multiple times and space where it can be experienced all the time. That is the big question for us. We are confounded.

It is not really answerable, but one can surmise that evolution or the desire to evolve must be connected to that state that is beyond the spiritual. I think of it, not so much as a steady state, but a state of awareness that has embedded in it this desire to explore. That is about as far as I can get with the question, and that is only surmising.

It is astonishing that we go from the physical to the spiritual. Correction. We are ultimately spiritual. What is beyond spiritual? That is an astonishing concept.

Why not ask the question?

What is beyond that? Now we are impressed. I am laughing with the 'light' (perhaps Han's expression for a new idea) because this is new to us. You have introduced something that we have not considered before. I am.... We find this very stimulating and enjoyable.

Well. It is something you can think about.

You are, perhaps, your species is more advanced than ours, with humility.

I think, throughout the Universe, there are always some variations. I might have better hearing. You might have better vision. I think it is just what we wind up with in our life journey. Not every human is interested in these philosophical questions. My soul sometimes travels in my dream state, and I get a beautiful sense of the lightness of flight.

We developed a humanoid type of body, and with this came the ability to develop tools and manipulate matter in the form of technology that eventually developed over a long period of time.

With this technology, along with our bold instinctual desire for flight, perhaps this is just what comes naturally. We were a happy accident.

(Therapist laughs.) You did well.

You were a happy accident. We are looking at each other and feeling sheepish. There is humility in this. I don't know why we are embarrassed.

And good humored, too.

Contact

We spoke a little bit about how the Zeta kin contact other sentient beings by various means. Some are obvious and some are not so obvious. Would you care to share what some of these less obvious contact methods might be?

When contacting other species, which we ourselves call kin in our relationship, it is often like communicating between humans and their species on Earth. For instance, the way you communicate with your dog, your ally. There are certain instinctual motivations as a species with whom we communicate. These are adjusted for in our communication with them. Some species are mostly humanoid in body type. What you would call humanoid. Sometimes they are not humanoid. As a result, based on the environment they live in, we can see how they would act and react to different notions and exchanges of communication. A dolphin may understand tools or apparatus. However, they would not necessarily see that as a commonality with us and technology. Yet we can learn from them also, particularly about their interaction with their environment and how they use it to communicate with other species in their environment. We notice that Steve is sensing a layer of a type of citrus scent, very dry, in a pristine environment. This might be the space that we are in. There are very soft, soothing tones and colors. There are pinks and amber colors that are very soft and soothing. The scent of our species is more of a.... It has an undertone of orange blossoms. Part of this is a scent that we have applied. It is a smell of a perfume, and that is one of the scents that is being communicated to you at this time. The space we are talking in is very comfortable. It is neither cold, nor warm. It is very comfortable. It is relaxed.

(Previously, Zetas have commented about their social skills coming from a desire to create a very harmonious atmosphere for social gathering.)

Are you receiving this Mary?

I am, indeed. If I understand you correctly, you are saying that your interactions with others would be determined by the available senses of each species. It might be through scent, perhaps color, perhaps vibration, and, of course, there is always telepathy.

These are many of the same senses, but with the added benefit of telepathy that humans do have, but it is often not fully awakened. There is the sense of telepathy as well as the other senses. Each species has a special emphasis on certain methods. Some are more predominant than others.

Is it the dolphin or the whale that uses echo location?

The sense of sound and sensory perceptions is more along those lines. Sound vibrations can be very sensual. At the same time, echo location can be a form of communication in interacting with the environment. It is used for communication and awareness. We find these species (toothed whales and dolphins) interpret this in their own way, knowing their surroundings. It is also used, in a sophisticated way, as a kind of thought exchange. It is a very beautiful way of interacting with their own species and the environment. It is like using Steve's allegory of ice cream with fudge sauce. There is a sense that it is served as nourishment, but there is an additional pleasure that it creates. Such forms of communication have multiple benefits and enjoyments.

We have talked about sound before, which is one of my favorite subjects, but I don't think we have talked about the sense of taste, in particular. Are there species where their sense of taste is their primary communication form or mode of interaction?

This would only be in conjunction with telepathy where you exchange what the senses perceive. Often, it would be with a one-way communication passed from one to another, only in relation to consuming food. One does not experience the taste of food alone in a telepathic society. We also exchange these experiences of taste with others, sharing in that enjoyment, or perhaps, non-enjoyment. If the taste is unpleasant, it can be communicated as well.

It is interesting that the commonality of taste can be a way of sharing a connection.

It has a connection, and much intellectual exchange can come from this exchange of taste. It is basic, but it does have a richness in how it can be interpreted by the mind, whether enjoyed or not.

Or if beings would take more time to enjoy it.

I remember your commenting about how human beings would sometimes stand outside and attempt to communicate with your craft by signaling with their flashlights and that, sometimes, the craft and the beings inside would respond as a form of learning, but also as a form of entertainment.

Entertainment?

Entertainment, I think is for the humans. Perhaps that is just the side effect.

Entertainment can be on both sides. Zestra has been speaking here, primarily. Han has been interjecting at times. The enjoyment comes from the happiness of contact made and the excitement that comes from the intellectual gratification that contact has been made on some level. A thought process has begun from that exchange of communication.

It is a great joy for human beings.

It is the excitement from the experience, learning that contact with a visitor has been made, and the thrill and the expectation and knowing that they are not alone, which they are not, that creates validation of their expectations. We can sense their emotional reaction from that distance and their delight. We feel the enjoyment from participating in that.

I am pleased to hear that.

It is like a wave, and we are waving back.

A gentle way to communicate.

We enjoy the exchange that comes from that, not only in reaching out, but the acknowledgement that is returned in that exchange. We use that as an opening for further contact, which develops from that initial point.

Yes. Not everyone here is afraid of contact.

And, they should not be. Our interpretation (of human's using flashlights to signal) is one of curiosity, excitement, and further curiosity. I will respect the interpretation of each species in how they perceive our contact. This is what we often observe from humans.

The Universe

We have talked about travel, of course. Using quantum mechanics, are there ever any accidents that happen in transit?

Gen, Zestra, and I (Han) are exchanging between the three of us. The use of quantum mechanics is a fact. Its existence, in and of itself, should not be blamed for any accident. It is only in the misunderstanding of facts. It is like a tool that should be used properly. If it is misused, then it will 'come back to bite you.' The understanding of it enables us to use it properly.

That is the second part of my question. Let us say that you are using it properly, and another species is using it improperly because they have failed to agree to Universal time, or they have failed to comprehend the full mechanics of it. In that case, might there be an accident?

There can always be accidents. They are caused by the inability to understand and use the tool of quantum mechanics properly. Just as with a human aircraft, they can use it improperly and crash. Adversely, they can use it properly, and it becomes a thing of joy and expression, an extension of their thoughts. We love to fly, too. Space travelers must have a respect for and understanding of the quantum physics and the time-travel relationship that they are engaging. When understood, it enables quantum travel to become an extension of their reality. It can serve them well if used correctly. This communication is becoming more philosophical, I suppose. Yet we are all in agreement with this. This is basic information. Zestra is the more practical one here, but I can be more philosophical, and Gen is also practical. The feminine-masculine becomes irrelevant in this discussion. Han can be the dreamer and the practical one. Zestra is more the practical one who can appreciate the romantic. Gen is also both.

There is room for everyone.

There is room for everyone. I am just outlining their perspectives.

Thank you. That is very helpful to me.

This is to help you to understand each personality.

Thank you. That does help.

I know that is something that will help you in interpreting our views.

Yes.

When a feminine one is very practical, it can be attractive. It is attractive to the others as well. It is not so much attractiveness as a description of how we are individually. It is just how one might view our personalities. The quantum physics that we use to engage in travel practices require a basic understanding of the forms of light. They have different, but similar meanings. I think I have shared the thought processes of each of us, which are the same and yet different. You are getting a mix of thoughts here.

And I am appreciating and enjoying it, too.

You are hearing not the individual, but a chorus, so you are hearing many thoughts at the same time.

I am very comfortable with a chorus.

We have exchanged with you with a number of us talking at the same time.

Open Session for Comments and Questions

I think I had better get ready to bring Steve back. I have so enjoyed having all of you here. Is there anything any of you wish to say in closing?

No. We just enjoyed sitting in the garden with you today, all four of us. Maybe there are five today.

You have, of course, visited many places and many cultures. Could you describe one or two cultures that you have come across that pleased and surprised you by their ways of thinking or perhaps their ways of doing things?

There are planets with oceans that often have intelligent life. Their social patterns also resemble more the patterns of land beings than ocean creatures, which often collect in schools of fish. These are pleasing to see, and it provides insights into how these patterns form, which are always engaging. There are also advanced forms of life, which are more energy than physical substance or pattern. These energies can be in the form of light and mass that moves freely throughout the Universe. They travel between the stars and the planets, both in clusters or individually. We have often encountered these, and there is often an exchange, a sense of respect, between one life form and another. They fascinate us. We wonder if these are forms of life that have evolved from a lower level or moved up to another dimension. Perhaps these are the spiritual beings upon which we often have commented. We marvel at their ability to freely travel anywhere they wish, using little expended energy. They seem to have an omnipresent connection on a scale that surpasses us within the society of our planet or our telepathic connection with other planets. They have an ability to move that distorts space and time.

Like inside a bubble or perhaps a distortion?

They are energies, and within these energies, there are bubbles, as you say, of consciousness, containing great wisdom and power. They can still use the material Universe to harness energies to interact with living and inanimate objects, collectively and with intelligence and purpose that is beyond the concept of most life forms. They exist on a level that is an inspiration to us. As advanced as we may seem to you, we are astonished by such beings. They do not exist on any particular planet. It is more diffuse throughout the dark matter that constitutes the Universe. They use the dark matter as some life forms use the seas to travel and to exist in a way appropriate for an advanced life.

Do you acknowledge each other when you meet?

Yes. Our contacts are possible, and they show a respect for our primitive means. They are not so advanced as to be detached from all existence. They appreciate our desire to connect with them and to be a part of them. In exchange with some of our more advanced leaders in our culture, these beings have given us insights that have created a spiritual structure to some elements of our society. These energies are difficult to define. There are many of them, and they abound in the Universe. We often wish to impress our consciousness with them so that we can experience their reality, and they can take us on their travels where they go to show us other realities, much like I am attempting to do here with you. They, in turn, reveal a door that we also endeavor to pass through and understand. They invite us. We invite them.

Do they travel interdimensionally?

Yes. Traveling in linear-time fashion is incompatible with this life form. They use interdimensional travel in much the same way as we do, but with a more ethereal form of pure energy. They are an energy of consciousness, a form of light, without matter.

Chapter Four

Session Thirty-Nine
April 15, 2012

(Through hypnotic suggestion, Steve is sent to work with Scala, Lent, Nicole, and others on a planet-scaping project in another dimension.)

Han, you are cordially invited to speak with us today

Hi, Mary. How are you?

Very well, thank you. I wasn't sure if you would have any difficulty getting through due to distractions in the home.

I was 'anxiously' waiting. When you invited, I stepped forward. Zestra is with me. Gen is not. Zestra can more than speak for Gen and me, too. After a shaky takeoff, we are here.

(Therapist laughs.) We all have to deal with different things.

I respect Steve's home. So that is behind us, and we are present and ready for our conversations.

I hope that you, Zestra, and Gen will ask questions from time-to-time. I am quite willing to tell you what I can.

That could happen.

Earth History

Not long ago, we spoke about how you shifted from having carbon-based bodies to a combination of carbon and silicon-based bodies. I was quite surprised to hear a geneticist talking about the possibility of creating human babies with a primarily silicon-based system. I don't understand why they would particularly want to do that. If they could, would that be of benefit to the human species?

Like our species, the human species cannot become a fully silicon-based being. There is a certain universality to being carbon-based. It provides a broader base from which life may thrive. However, there are certain elements of a silicon base, which are compatible with carbon-based life. The role of the silicon-base would be to extend the elements or properties of carbon-based life.

Life extension?

In part life extension, but also to augment certain limited elements of carbon-based life. Certain parts of our body structure are improved with silicon-based components. There is actually a hybrid type of prescription where both carbon and silicon are intermeshed for an optimum use of both. Our origins are also carbon-based. Now we have augmented parts of our biology with silicon-based elements to extend our lives, making life more convenient for long-distance space travel, and making our biology more adaptable for living in a wider range of planetary environments.

You did that when you became spacefarers?

Yes. Zestra is the primary one sharing now. It is not very different than Han's thoughts. Our thoughts and actions are compatible.

Thank you. That does help me to understand why Earth scientists would want to reach in that direction.

It is to embellish the carbon components with certain silicon-based components to enhance expansion beyond normal expectations of carbon-based life. In the Zeta kin, the original parts are carbon, but some silicon bases have been introduced to embellish the carbon component.

As you know, human beings have a pineal gland, which, in humans, can under certain circumstances, produce hallucinations and may trigger out-of-body experiences. I read that the pineal gland, in ancient history, functioned as a 'third eye.'

The pineal gland is used for that role and is often a part of the components used to help awaken the telepathic abilities in humans to enable communication with us and many other species. It is what we have discovered in the humans that we call Catherine and David (participants of the Serpo Project who now reside with the Zetas). It is one tool that is used to study and expand the connection between the Earth kin and our kin.

Did you have a pineal gland at one time? If so, do you still have one?

What we call such has been expanded in its role. It has a greater capacity. We are working to expand this in the human species so that our species and yours have a harmony in such communication.

Zeta Reticuli History

In discussing telepathy, you indicated that we each have a 'reach of the mind.' Do you see limits on how far your mind can reach?

Our species has a limit. It is overlapped among others of our same species or 'kin.' It is quite universal within our species. But there is still a limit, a physical distance. We have certain members of our species who have concentrated and focused to extend it further, even into a quantum state, thus communicating with species on other planets. We often rely on these members of our species who have become ambassadors, such as Han and, to some extent, myself (Zestra). We have trained ourselves with special training, much like your monks and, to a degree, much like yourself, Mary. We use this to reach out beyond our normal species. It is also used to a greater degree among ourselves, to overlap our thoughts and emotions with others of our species to extend understanding. This is our normal mode of communication. Some express through other means beyond their normal telepathic abilities. Even among our own species, we have to form relationships and understanding to enable us to extend to other worlds by a quantum physics means. This form of communication between worlds does not require physical transportation, though we do have that ability, too.

It seems that this free exchange in communication across worlds also facilitates commerce.

Quantum physics provides this, and it does save time and space by not requiring a physical transportation of communication. I hear your ally. *(My dog is whimpering.)*

She is whimpering in the other bedroom at something unseen. She does it intermittently.

Do you wish to defer attention to her?

No. I will make sure the door is open, if you will excuse me for a moment.

Her paradigm is important as well, so please attend to her so that she is comfortable as well. I know your ally is important to you. Her needs are important to us as well. I have visited your home and know that she is a lovely being. I have fondness for her. *(I find this an astounding level of thoughtfulness on Han's part.)*

When I was planting vegetables in the garden today, I thought of you because I know that most of you plant some food for yourselves. I enjoy that we have that in common.

We grow many plant species for the pleasure of the beauty they provide. We plant both non-edible and edible plants. We are often like the English of your species who also enjoy gardening, or the Japanese, or other cultures of your world, which enjoy it for its own sake. We have much in common and have a harmony that we share together.

You know how biological beings can have vestiges remaining from evolution. We have such vestiges as our appendix and our tonsils.

Our legacies. Legacies of what we were and the legacies of what we are to become. That is something we can appreciate together, in our own respectful way.

Yes. Do you have vestiges remaining in your physical bodies that you no longer use?

The vestiges of our sexual past have diminished due to our new form of biological reproduction. Parts of our senses, our ears, our vocal cords, the sounds we utter and the purposes for them, have become obsolete due to the extension of parts of our minds that now make communication more efficient. Somehow, music has changed for us. We have spoken of music. Our minds seem more connected to music, yet the sound of it, and the voices that utter that music have changed.

This is a good example.

That is one example. Certain body functions have diminished and others less so, as we evolve. In a way, we are saddened by the loss of these, and yet we know that it is just part of what we are evolving into. We look to the higher parts of our minds with hope and astonishment as we trust what we are becoming, as you also trust. The passions of the past and the thoughts of what we are to become in the future provide that peace in whatever we are to be. We hold in trust whatever our evolution may be. We trust in and place hope in some higher powers, as we wonder together and look to the stars.

I am just so grateful for having lived through so many interesting events. Life can always become even more interesting.

I have shared the thoughts coming from us with the sense of hope and vulnerability, knowing that what we share comes and goes in ultimate good. In the infinity of space, we are just a continuity and, at the same time, just a glimmer of the light that we both a part.

Who is…. What…. (multiple thoughts coming through all at once.) What will be is accepted with trust.

Thank you for expressing it in such a poetic way. It does express the intensity of the feeling behind it.

Zestra: This holds for female-to-female, and the male, too.

I remember the term nano technology from some years ago. Could you tell me about the progression of nano technology here on Earth, and how Earth scientists think they might be using it in the future?

There is much to be gained from this. In your health and life extension research, which you call life extension, due to help in medicine, know that it is just one new avenue. It will provide health and continuance. It will provide a continuance and a merging of technology and biology where, at present, the two are considered separate. This merging will provide distinguishing characteristics between carbon-based and silicon-base life. This is not one to be feared. This is just one that is a natural extension. You are learning the early primitive forms of this, and nano technology will provide the basis for this.

Why does your craft come to my mind when I think of nano technology?

There is often a psychotronic connection between our spacecraft and your biology and our biology. There is a connection there that will provide an understanding between the artificial intelligence of our craft and your and our biology connecting and communicating between them.

Thank you.

Spirituality

When I want to visit the spirit world, I do not look outside of myself. I go deeply into myself through a meditative state. The information flows to me somewhere between the subconscious mind and what we call the 'higher self.' Do your minds also need to go through a gateway such as this?

We do. (Zesta is speaking.) Han is with me on this. We are all connected. In some ways, Han and I, from a human perspective, would be interpreted and misinterpreted as lovers, but we are connected in many ways, as we would be with you. We stand in the infinite of the smaller Universe with the infinite of the larger Universe. We also stand in another way, the way the 'third eye' as you refer to it, might refer to the infinite spiritual Universe. In a way, we stand at that doorway.

Thank you for sharing that.

You are a part of our Universe. We enjoy this, as we know we are a part of your Universe, too, with respect.

Thank you. It makes it a much more dynamic and interesting place to be.

It spins around us.

Yes.

It does, I am sure.

I had an experience with a being called Metatron, when I was doing deep spiritual practices.

This was your person (Djwhal Khul) from Tibet?

This was a different one.

During my meditation, Metatron, (Enoch, ancestor of Noah, transformed into an angel) stepped forward. His presence created a highly magnetized field, and I could feel that it was quite strong. He then took me to a place that was filled with what seemed to be countless photons, floating and interacting consciously with one another. Each photon had full intelligence, and I was aware that my consciousness was one of those tiny points of light. I felt very unencumbered. I suspect that who we are can be condensed to very small points, at least as small as a photon, maybe smaller. Yet who we are remains intact. Do you have any thoughts?

It is not an accident that the sense of light, photons, are a part of this process. One can feel like a pinpoint of light and, at the same time, the source of that light, and at the same time, reflecting light, all at once. It is one of both a physical and a spiritual state and, beyond spiritual, there is a connection with the Universe. It is one that we have often achieved and are endeavoring to know more.

It was curious that there was such a highly magnetized field around this process.

It is characteristic of this process. It is one of *focus* and *expansion* at the same time.

I think he may have just wanted me to have the experience because a number of us were communication with him as a group. I think it was his way of sharing.

He wanted you to share in that emotion you call pleasure, to create curiosity, and to know the peace and wellbeing that comes from this.

I could feel him as being incredibly strong, complex, and having very good energy.

Contact

My first question today is about Project Serpo, the project of moving humans off planet and moving Zeta kin onto Earth as part of an experimental exchange.

Yes. We have discussed this earlier.

Yes. I was wondering if it was possible for us to talk to one of these human 'immigrants' to your planet. In particular, I would like to talk to one who has chosen to live among you, rather than staying within the human grouping. Are there very many remaining who might be available?

There are a few. They are older now. We have arranged to extend their lives so what is considered 'old' is not now 'old' for them. It is extending the normal human life span to a degree.

Are any of them telepathic?

They have acquired some skills, and the abilities that have always been within them have been awakened. We have been training them to focus, to connect. Also, the training includes developing the ability to ignore, so they do not hear the din of other thoughts plugging in. It requires a kind of relearning to focus thoughts to hear a voice in a crowd and to concentrate on this or a conversation. This is Zestra, and I agree with this. There is a person named David, who is a human and also a female named Catherine. They are two human residents on our planet, and they are happy to speak with you. They are a human couple who were part of the original group who have formed their human relationship and now live together among us. We give them the space to interact with us, yet allow them to be, to a certain degree, alone in this human relationship and connected to us at the same time. They are not segregated from us. It is impossible to do this completely, but we allow them their own space within the confines of being human and forming their own private relationships. I am trying to communicate that we have given the humans their own home. Han is agreeing with this.

The fact that they have chosen to live among you speaks to me of their ability to be curious, adventurous, and to live outside what is the normal comfort zone.

They found each other and formed a human relationship, which we study. At the same time, we have provided comfort for them. They are both adventurous and willing to be together. We are with them, and they know that, if they wish to return to Earth, this can be done. They are happy together and have given birth to human children. They are alone and yet not alone. They are enjoying the blessings, bounty, and beauty of our planet. They feel that they are lucky. They have lost a lot of their own human ties and relationships with the past, but they have formed new ones between themselves and with our species. They feel, in a way, that they are in an idyllic situation. There are ones with whom they can interact, and they are deprived of nothing.

With humans, if we are in an unfamiliar land, and we should be walking down the street and suddenly encounter a wildflower that we were familiar with, it brings back all kinds of memories and emotions. It also has a kind of anchoring or emotional grounding effect. Familiar things are important. How do these two humans find their comfort in familiar things that reinforce their origins?

Your client (Steve) is sensing the smell of flowers, a clear, clean atmosphere, and sunshine. Through Zestra and Han, the impressions of these humans are coming through. The happy thing is, although they live on an alien world, there are pleasant things that are not unfamiliar. There are many similar things, some of which have been provided for them. Due to the transplantation of many species of plants and animals, they are surrounded by plants and animals in an environment that is not dissimilar from the planet that they remember.

Are there animals around them, perhaps like cats?

They live near an ocean, and they can see animals that live in the oceans and on land, including elephants. There are elephants. There are cats, dogs, and birds. These have been permitted because they do not destroy the ones that are interconnected in this ecology. Now, humans are also a part of this ecology. I see green rolling hills and flowers. Also, plants that have been provided. They are arranged according to our own style of bio-connection in orderly parks. The planet is arranged so most things are arranged in idyllic garden settings. It looks very idyllic, one that humans would find very attractive to them, as well as to us. There is an order, but there are areas of our planet that are left to grow wild, in addition to the arranged gardens we enjoy. They also enjoy a botanical garden. They jokingly refer to this as their Adam and Eve Garden. We know of this mythology, yet they do not live in a menagerie. This is their reality as it is our reality.

How do they travel around? Do they walk most places?

There is walking. There is also a way for them to transport themselves telepathically, and they can also create, in a physical sense, everything that they need. Food is grown, but they also have the ability to materially create food, as well as other things that they require. We have an ability to do this. If they require a ship to transport themselves, it can be provided. If they wish to have a home, they can imagine it, and it can be physically provided for them. *(Humans imagine their needs, and the Zeta kin create a reasonable physical facsimile of what is imagined.)*

It is rather like a few descriptions I have heard of heaven. (Therapist laughs.)

It is not heaven. Perhaps it appears so to some. You can imagine why they do not wish to return to Earth.

Yes.

There is a beauty in this. It also provides us a way to study humans in a kind of idyllic setting. They have children, several of them, which they reproduced through their normal means, a means that we have abandoned or replaced. It is a study of human reproduction and the relationships that form as a result. These children will also reproduce.

The children will be different from their parents. How do you expect them to evolve over the next couple of hundred years?

They will populate our planet. In our species, population is limited. In time, the humans may also need to be transplanted. They may even wish to return to their Earth. They have the freedom to do this. It is, in a way, an experiment. It is also a dwelling space for them. It is not an experiment, in that sense. They are guests and also permanent residents with all of the conditions that are bestowed upon our own domestic residents. They are the same.

How do you feel about humans who are born on your planet? Does that mean that your planet now becomes their home planet?

Catherine and David are considered to be of Earth, but their human children, born on Zeta Reticuli, are residents. This is their home planet. The stars that they see are seen from a different vantage point than one sees them from Earth. The constellations appear different to them. They see two stars that provide light and heat for us. Their length of day is different – 2.3 times that of Earth days. Their year is different. Their spring and fall are different. If they wish, they can return to Earth. Yet there is no desire to leave. To them, Zeta Reticuli is now their home.

I am very pleased that they have at least some familiar things around them so that they have that foundation, but I am also quite pleased to learn that they have children.

They are allowed to follow their normal, natural life processes that are part of the human experience. They are free to experience and enjoy this as well as experiencing the difficulties of being human. Their natural life spans have been extended because this is part of the natural process on our planet. The surroundings have been provided for them. Some of the things familiar things to them are also familiar to us. The provisions given to them are also provided to us to enjoy.

Are you in contact with them right now?

I can make contact. David and Catherine are available. We are accessing them. They are here for you.

I would like to know what it is that they hope for the future of their children in terms of what they might do with their lives.

They are being contacted, and they are greeting you. They are unexpectedly surprised to be contacted by another human from their home world. Their children are amused. They feel they are being contacted by an alien, but they know you are not an alien, just an off-world being. They see you as the same, but also see you as a human that they now need to become acquainted with. Catherine gives you a warm touch and David, also. The children are surrounding you now. They are curious about you. Your client's voice is changing. There is a warm connection, and the thoughts of the children say, 'So this is what a human is like.' The parents are telling their children, 'She is like us, another one.' The children look the same. They are very familiar with Zeta Reticuli kin. It is kind of wondrous, different, but the same. They all wish to share how happy they are here on Zeta Reticuli: 'We (the children) have heard of Earth, and we hope it is still the same, and that it has not changed too much. We hope that the people of Earth are okay. We sense that they are. Please know that we are happy here, too.'

The reason that I asked that question about what the parents might like for their children's future is because the children are in a completely different situation. What a parent might like here on Earth for their children might be quite different.

They (the children) have relationships with sentient life forms. Their telepathic abilities they have acquired from birth. They know that they are different from the domestic species. They are expected to interact with them. Both are humanoid, but there are differences. Other humans can be provided for them if they wish to form relationships as they mature. They know that our species (Zestra is speaking) accepts that there are certain developmental actions that require human contact for them to reproduce, and they will meet other humans, if they wish. At the same time, they have a connection with us that is quite developed, and they are integrated into our lives. They know that they are protected and safe. In addition to this relationship with us, they are also given their own life opportunities. They will also have all of the knowledge and advantages that we can provide to them. They are taught how to live and conduct themselves in our society, with all of the blessings of an education that includes extensive knowledge of the Universe. They know more about the ways of the Universe than human scientists. They do not have the restraints of being limited in what knowledge can be provided. Because of the consequences of living in our culture, they have been given the full knowledge of the wonders of Universe as we know them.

Here, (on Earth) to live a good life, one needs to perform some kind of service for those around them.

They help. They help us to understand humans. They, in turn, help humans to understanding us. At the same time, they are also forming and contributing to the social work of our world. Technology and sciences provide a pathway to service. Sometimes it is in food preparation. Other times, they are involved in the exploration of the sciences and the technology of travel. Their lives are occupied and fully engaged in the human desire to feel fulfilled in what you call a 'career path.' Male and female are both equally engaged in work, both for themselves in their research. They also feel fulfilled in their roles socially and biologically.

I am very pleased to hear that their lives are full.

They have been a success in the Serpo Project. Not many have accepted this (remained on Zeta Reticuli) but have returned to Earth. Those who remain are free to stay. They have been a success in the exchange that was beyond anyone's expectations.

I will continue to think of them from time-to-time.

Know that they are loved and accepted and that their world is remembered. Know that those who have returned to Earth have fond memories of our world as well. Those humans who have remained with us are accessible in that they have access to other worlds to visit, if they wish.

I thought that would be the case.

Their children will also have the same success.

I think that is a good path. Thank you very much for the contact with David and Catherine and their children.

I am being flooded with thoughts. Let us just say that they are also very thankful to you for the extension of contact with them. They are profoundly moved that they are thought of, and that they are remembered, and that they are thought of by you. Know that you are part of something special in being thought of and remembered and they want you to know that they are grateful they are not forgotten. They will never forget you.

I remember the others, too, even though I have not met them.

For us, Zestra, Han, and Gen, as well as David, Catherine, and their children, today is a special day for them by being remembered. I (Han) am sensing their tears of emotion coming in, perhaps tears of relief.

Please tell them that they only need to give a thought. We can always make contact.

It is like getting a long-distance phone call from home and being brought back. Know that they are happy in being brought back to the time they remember. They remember that they are all doing their part, no matter where they are, even far away.

Please create for them a bouquet of California poppies.

And know that...... (Long sigh.) And that is the way it is.

Other

I learned the other day that the human cell has a finite number of times it can replicate – approximately 50 times, before it no longer functions. As cells lose their ability to regenerate, we age. Can you tell me how many times one of your own body cells would normally regenerate?

Our normal cells can regenerate as many as 700 times.

When they stop regenerating, does that tend to happen in a cluster or just a random cell now and then?

It depends on the function of the cell. Our brain cells do not reproduce so much. With our other cells, some are exposed to the outside environment and reproduce more often. It depends upon the role or task of the cell within our bodies. Some of the humans we have just visited have had their body cells modified to extend their lives. Though the human brain cells do not really reproduce, due to the extension of their other cells, in order to maintain compatibility, their brain cells have been modified. Our bodies live much longer. We spoke of a thousand or more years. This is part of normal life. When you consider the orbits or the way you measure time, our lives are quite long. The humans on Zeta Reticuli have extended lives to adapt to our cycles of time, including seasons, which are much longer than seasons on Planet Earth. The period of cycles and seasons are different, but we have accounted for this.

Open Session for Comments or Questions

You know our friend George.

Yes. He is one we regard with respect and affection.

Do you think he would be agreeable with making telepathic contact when Steve, Doc and I are with him?

I do not see why not.

I am hoping he will be comfortable with that.

I have connected with his thoughts. If he chooses to do so, he will. If he doesn't, we understand, but we hope he will. I do not see any problem with this.

Is there anything that any of you wish to add before I bring the session to a close?

No. Yours is always a curious question. So much has been exchanged, and yet you ask, 'Is there more?'

(Therapist laughs)

We look at each other. We smile at all that is done. Yet there is more to be done. There will be more in the future. Know that the portion that has been done today is sufficient for today.

Yes.

With a flow of happiness from all who have been touched today, David, Catherine, and the children, Gen, and me (Han), Zestra, Steve, and you, Mary.

It is a quite a good group.

It is a family and much has been shared today, and with this we leave, knowing that we can come back any time.

Thank you, and I send you a fond adieu and Namaste.

This comes with a burst of emotion and energy from the humans that I did not expect. They just wanted to share so much. They were not lonely or unhappy. They may never have this opportunity again, but they just wanted to share that with you. On that point, we say goodbye for now.

Until we talk again.

Session Forty
April 29, 2012

Steve is sent to work with Scala, Lent, Nicole, and others, on a planet-scaping project in another dimension. (This is a hypnotherapy technique to remove the attention of the client from this exchange.)

Han is cordially invited to speak with us today.

Hello, Mary. This is Han. Zestra is also here. There are two for today's session.

Hello, Zestra.

Hello, Mary. It is nice to talk with you again.

Thank you, and thank you for allowing us our little Sunday trip. We were not able to set our usual Sunday meeting last week, but we knew that you would understand, and you knew what we were doing.

It is not a problem. I know that humans sometimes need a mental break. That is quite satisfactory.

Thank you for your understanding.

Earth History

In human folklore, there are references to blue-skinned beings having visited this planet. Are you aware of any beings fitting this description?

Yes. There are species of small beings, very short, squat in appearance. They often attend with other species. There have been a number of encounters. Whitley Strieber is an experiencer who has encountered such beings, and others have, too. They often appear as entities that have been known in early human legends. They have been misidentified as deities when, in fact, they are other beings.

I have seen drawings or paintings of them from India and Indochina. I suspected that the encounters were true.

I have an image in my mind of their appearance. I am placing an image in Steve's mind. They look rather troll-like in appearance. This has translated into human legend and in the areas that you have mentioned.

These came from off-planet or did they come from this planet?

They are from off-planet. They are not from under the surface of Earth, as many humans have speculated.

Thank you. I am sure the blue coloring served some evolutionary purpose for them.

If I may, I would like to speak to you about domestic cats and how they purr. The vibration of their purring is about 25 vibrations per second. I mention this because studies have shown that cat owners have had their heart problems reduced by 40 percent by virtue of being in the vibration of a purring of the cat. Are you familiar with this?

The harmonic frequency is very emotionally healing to humans, and it has a relaxing effect, internally, externally, and spiritually. It tends to have a positive effect on certain autonomic reflexes in the human body, the heart, particularly. It would make sense that this is so.

When I am around the purring, it is almost tactile, which sounds strange. I sense the purring in more than one way.

Do you have the same physical response with the species you companion with called dogs?

I don't have the same visceral response. A dog is very comforting, but a cat is different. It is the purring itself that connects with something in my brain, a kind of entrainment. Maybe there is even intention as part of the purring effect coming from the cat.

There is an electrical energy in one form that is exported to those surrounding it. It is an inclusive vibration that is very seductive and very beneficial.

Do you think that this is something that cats evolved to create a kind of harmony with those of their own species?

Yes. Cats are both social creatures and individuals. They are loners, yet they are very social as well. They use scent and other forms of marking to help distinguish certain social patterns in their lives, including territory and companion selection.

Yes. I guess a number of animals use their scents in different ways. Probably you have heard the expression that a cat has nine lives. A veterinarian talked on the radio about his experiences with cats. He said that he has experienced, more than a hundred times, cats being brought into him ready for death and then making full recoveries overnight. He doesn't know what to make of the amazing recuperative power of cats.

Ultimately, they do physically die, but, before this, the reasons for their recuperation may be many. Some may be well understood. Others are less so. By surrounding themselves with their own kind in a setting that is positive and reassuring, it can bring out a mysterious energy that can be restorative. It comes from within and from surrounding external sources in the proximity. It is very mysterious.

We recall from Steve's memory that he rescued an elderly cat. It was very ill and, for a time, the cat rebounded. Eventually the kidney disease overtook the animal, and it died. However, the animal did not die alone, and the animal was grateful for the care it was given in its final days. The energy that enabled the cat to temporarily rebound was similar to what you described. Eventually, physiologically, his health became debilitating, and the cat succumbed, but his life was extended for a brief time. In Steve's memories, he had never heard an animal expel its last breath before, and it was very new to him. There was no inhale following an exhale. It is a memory that is very vivid in his mind, and the strange cat was taken to an emergency room. There the veterinarians were suspicious, but they soon understood why the cat was brought there.

What were they suspicious about?

That maybe the cat had been abused in some fashion, but it showed no physical signs of abuse, and this was not the case.

A rather harsh assumption.

It was just a precaution. Steve sensed this and was annoyed by it, but it was not the case. In either case, it does not matter. It was just that he had never experienced the death of another in such a vivid and personal fashion. He wished for the old cat to have a comfortable resting place so that it could die peacefully in a comfortable and loving setting. This is a trait that we find appealing in him, as I know others do, too. This is why, among other reasons, he is facilitating our communication.

He has his own kind of goodness and compassion. On the reverse side of this loss experience, I had adopted a semi-wild cat that started crying and acting restless. It came over and climbed onto my lap. I did not know what was the matter with it. Then the cat kind of stiffened its body. It decided to give birth in my lap. That was a complete surprise.

Han and I are showing astonishing surprise. If you could see our faces, our mouths are open, the way humans show surprise. We share this facial expression.

(Therapist laughs.)

How many offspring occurred?

I think I remember at least two. It has been a long time.

I hope your clothes were not too soiled.

(Therapist continues laughing.) When one lives on a farm, one does not care about one's clothing.

Then it was a setting appropriate to the mental framework of the situation.

Yes.

If I may, I would like to ask about major floods on Earth, and I am sure there have been many. There is a belief in some circles that some floods were intentionally caused by off-planet beings to remove the life forms that they had altered after finding that they were unhappy with the results. Is this something that happens around the Universe as beings dabble with genetics?

Erasing a 'chalk board?' It is one technique that is a temporary removal without destroying the planet. Often a planet can be destroyed by just removing its sun. This is not discussed. It is too terrible to think about, but it does exist. It is possible. You have had many floods for many reasons. There have been brief life stages, and the thawing and consequent flooding have often caused adjustments to the evolutionary process on your world. There was a time when your desert regions were very plentiful with plant life. That is no longer so. Some of this is due to climate changes and meteor impacts, which have caused temporary overcast of debris in your atmosphere. This adjusted the environment to lessen life.

Has it ever happened on your home planet, an intentional reduction of life en masse by another species?

By another species?

Yes.

Maybe in our ancient history, long before we became aware. This could have happened. We have an awareness of the sky. It is part of our consciousness, so the possibility of such activity is not foreign to us. Your world is inward looking, not outward. Much of our existence and environment is so much determined by what is in the sky, as well as on the ground.

Do you mean the area right around your planet or do you mean beyond that?

Beyond that. You (Earthlings) tend to view those things that affect you as coming from your planet. To a great extent that is true, but there are other effects from external, off-world factors that affect life on your world, too. For example: meteors, cosmic rays, and other energies, and other civilizations that could be contacting you. It is an Earth-centered viewpoint that you have, much like your early astronomers on your planet who thought your world was the center of the Universe. Now you know that is not so, but your mental state still feels you are the center of the Universe. (This is changing with the discovery of exoplanets.) As we know from our communication, you are not the center, but part of a larger family to be discovered. With that comes an expanded consciousness and responsibilities. These responsibilities should not necessarily be stressful to you. It might require relinquishing old patterns and beliefs.

I don't mind.

Some may not mind. Some may mind, but that is your reality, and you will start to incorporate new facts.

I look forward to it.

Humans cling to a rough stone and give it great value. It may be exchanged for a polished diamond, if you wished to let go of the stone in your hand to make room for a better stone. Maybe in time, that will come to pass.

In this particular life that I am living now, it has been a series of building beliefs and discovering new information, expanding beliefs, and letting go of beliefs. It just continues on as more and more comes to light. I am quite comfortable with that and quite excited about it, too.

Knowledge should not be feared, just incorporated into one's existence.

Zeta Reticuli History

Since we last talked, have you met any new sentient beings who particularly caught your attention?

We continue to explore the energy entities in a far-off region, exploring them and, in turn, their exploring us. There is interaction going on in that area of space. This is well known among the community of minds, and they are being welcomed. The contact is a very positive one as we learn to interact with them in a new way. It is a challenging concept for our species. As you are exploring us, we are exploring them, and questions often surface.

Who made the initial contact in this case?

As best we understand, it was both species. The other species is more pure energy and, as such, it is not so much traveling to a destination as it is traveling to another frame of consciousness within each of us. They occupy another location in both in space and in consciousness.

I would think that you would find it very gratifying to get to know this new group. Of course, seeking out new life forms is your life's work.

It is a new frontier, and it represents a species that has evolved to another state of existence.

I am not surprised by that. I feel grateful to learn of their existence through you. Of course, I have learned that sentient beings come in all different types of physical forms. The interesting point is that, as they evolve, make contact, and move from the physical to pure energy states, they seem to have a shared understanding and respect for life forms. At least that is the impression I have.

Energy becomes even more universal in where this energy dwells and emerges from. It does lend itself to that commonality.

When I reflect on us coming from the same form of subatomic particle, and that we move back into that state upon death, we never lost the commonality. We simply have intervening lives.

Human and other sentient consciousness emerge from very base forms of elements. Ultimately, we are learning that such energy, sentient energy, evolves into pure energy, and it begins to return to the appearance of those simple basic elements once again.

Yes. Interesting.

It is like a whole, new perspective. At the same time, this appears to be deceptively simple. They are more advanced but, collectively, they appear to be a very basic form. It could be easily misunderstood.

Could you make a slight adjustment to his microphone?

Yes.

There was a little too much sound kickback. Thank you.

Referring to the 'dust-to-dust' or subatomic particle-to-subatomic particle concept, I need to mull over how the intervening life experience can still be retained and stored in very small units. (Therapist shifts from these weighty thoughts to Zeta biology.)

You talked of using a kind of perfume sometimes on your bodies. I think this is part of your normal regimen to provide comfortable and appealing conditions for those around you. Does perfume, as you call it, serve more than one function for you?

I do not recall us referring to such scents, but we do.

Orange blossoms was the scent you mentioned.

Yes. That is true. This comes from some of the surrounding flowers and fruit in the proximity, but we also impart that on ourselves. However, it does not have the purpose of reproduction. It is merely to add the fragrance that we enjoy and carry it with us. Sometimes we do this. It does not, as it does in human circumstances, serve as an unconscious chemical signal for reproductive purposes. We use it for the pleasure of the scent and as a way of carrying the scent with us. We sometimes apply this to our bodies to enjoy it.

It was curious because I thought it might work as both a deodorant and a scent, but it does not sound like it is needed for that purpose.

Sometimes it can have therapeutic purposes.

Oh, yes.

Some fragrances are used. The use of fragrance is from our past. Perhaps it is a fragment of our once more prolific vanity.

(Therapist laughs.)

It has diminished along with our emotions.

I am thinking the term 'essential oils' might apply for therapeutic purposes.

Yes. This is not dissimilar from our uses. We are careful not to apply it to our heads. We tend to apply it to our arms and back, but not so much to our heads because the acids, or the alkali in fruit, depending upon its level, can irritate our eyes. Compared to your eyes, our eyes are more prominent, and it can be quite irritating, but applied to other areas, this has become accepted. It is a custom for us.

It sounds like it would create a nice experience to be in a room with several individuals who are using this 'perfume.'

Our sense of smell has not diminished much over time. Our hearing and vocal skills have diminished. Our sense of smell, as well as our other senses, are still quite active and effective. We can distinguish individual scents when the fruit essence is applied to the body chemistry, and we can often distinguish beyond with our telepathy, the individual scents of our companions in a large space. It is just one of the things we enjoy, among others, in our social interaction. This is applied by both the males and the females.

It is interesting that sentient beings, even after having been evolved for a very long time, still have strong connections to the plants around them.

This is perhaps a subconscious expression of that.

Yes. I know I feel very good around living plants and trees.

It is like lying down on a bed of flowers, and it is a scent that is pleasing to us and is enjoyed by many. These are personal things, but our concept of 'personal' may seem different. I think, between us, it is not personal. Perhaps it is just revealing another part of ourselves to another species that makes it seem so, but it should not be so in our case.

I see it as another interesting aspect of how you live life.

It is a more private part. It is not intimate since it is commonly shared by all in our population and culture. It may seem strange sharing this with another species, but it is not. In this sharing, I am just stretching myself as well. We are sharing things, which are, in some cases, unprecedented, and we are taking you by the hand into the center of our culture so that you can share in this. It is with a welcome that this is accomplished.

I thank you very much for your willingness to do this.

I was wondering if you could tell me something about a species that we call the Ebens?

Yes. We are much like the Ebens. They are grey, but with what humans might consider human-like features.

George (George Lobuono, author of 'Alien Mind') thought you might be something like them.

We would be considered cousins with another advanced species. We consider ourselves more advanced, by an order of magnitude than a typical grey, as you refer to them, though nothing is typical. George has another term for the advanced species, the hyperversals, with which we are more akin. We are even closer to the hyperversals than we are to the standard 'grey.'

I have the impression that there are a number of different kinds of beings coming out of Zeta Reticuli.

Yes. That is an accurate assessment. We are more an advanced version of the Ebens. We are not like his description of what he calls the Verdant, which we rather distain. We feel we are more advanced. Biologically, we are cousins, but socially and culturally, we are separate, both in a moral context as well as a social context. We do not often enjoy the presence of Verdants, though they do need to co-exist. We do not appreciate what the Verdants do. They are frequently involved in illegal practices, including human abductions and the abductions of other species as well. Their self-image includes pride and arrogance, and they feel they can take what they wish. However, we do not follow those practices. We understand and respect life forms, whether more advanced or less advanced than us. We are more traveled and have acquired this attitude from our many contacts with many species.

What have the children of David and Catherine (human offspring living with the Zetas) been taught about Zeta social niceties, and what would be considered a social faux pas?

Faux pas?

Errors in good manners.

This is a new expression. We are looking at each other. We are grinning in our thoughts. It is a fascinating word. It is debatable whether it is a realistic experiment (very long-term observation of humans in a natural setting). Since there are other humans of the SERPO group that are beyond in numbers, other than Catherine, David, and their children, there are also other human communities in various places. Is it a realistic observation of human behavior because the environment has been changed? They live in a new environment where there is little human contact with others of their population, but contact with our population. Is this a realistic experiment that we are observing? We have changed the expectations so that the purpose is for them to be successful as a family unit and to provide them with a satisfactory and meaningful life. That is foremost. It is just what we have to realistically expect from our experiment.

The offspring no longer have certain social constraints. In a way, they have been imparted with certain social patterns of polite inter-action, but they do not have certain laws or social constraints that perhaps other humans on your home world may have. It is like living on a tropical paradise island alone or in a small group where certain former patterns of social acceptance have probably become

Relaxed, and others have perhaps been modified to fit the current situation. The one thing we have noticed about humans is their great adaptability. That is one trait that has been studied, as well as their mental capacity to adjust to their different environments. By some standards, they would be considered to be living a luxurious life, a successful life in a biological way.

The one thing that they are perhaps not experiencing is what happens in evolution when you have to face a certain amount of harshness in life. Although, I think you said that you left them to experience both the good and the bad. Human evolution brings times of chaos and times of peace.

Chaos is a common reality in the Universe. They must face the same chaos that we are facing. The exception is limited to our smoothing their way as they are introduced into a new environment. We do this so they will have an equal chance as we do. We have allowed for that. The allowances have been very small. There is little that they face that we do not also face, as well as experiencing happiness.

Then they will continue to evolve. One of the things that I see out of this experiment is that they have been given an opportunity to show what humans do under optimum conditions: good health, pleasant environment, loving beings around them, and access to knowledge.

They have been given access to tremendous knowledge, beyond anything any earth school would ever provide. Also, they have the ability to travel.

Oh yes, travel is a great educator.

A field trip takes on a whole new context in our culture. This is a good argument of the opposing view that we often have. Their existence on our planet is not in question. It is only our discussions that we study. As we observe the human existence here (on Reticuli I) these topics and issue arise. It is part of why we find it interesting and fascinating.

The same, of course, applies for the merged beings (hybrids) from your area who have been placed here among Earth kin, except for the advantages they have brought with them. It provides a different perspective on adaptation.

The merged beings have had less to adapt to because of their own existing abilities. It is almost a very easy walk for them by comparison. The others on our home planet are not experiencing difficulty, but the Zeta kin merged beings on your world, Earth, are surviving and flourishing, but with a secretive awareness. It is not that they are a secret, but the humans have made them so because they have concealed them from the rest of your population.

My hope is that it will not be required in the future, that they can live openly. I do not know when that might be.

That would depend on your species, but they feel, in a way, like an advanced student in a class where the efforts are more trying to promote stimulation and challenging goals to occupy themselves. They hope the humans will come to understanding them, especially as they demonstrate ways that they can make life better and more comfortable for others and themselves on Earth. Each new revelation can be so wondrous that what is taken for granted with us, the primitive-like mind may find shatteringly profound. The ability to displace gravity, for example, which your species still does not understand, is something that is part of our physics, and it creates the impression of something 'magical.' Yet it is mere physics. It is the same physics that applies everywhere in the Universe. To the local culture (Earth), it can seem rather wondrous. It is not. It is like introducing primitives to an airplane or a radio or a motion picture.

The initial wonderment.

The initial wonderment is profound, yet these are physical realities. It has effects on a primitive culture, like when a primitive culture sees a motion picture of an on-coming train. Their inability to understand a two-dimensional representation makes them run out of the way of the misperceived approaching train. This is an example of how your culture might perceive revelations that are from our species interacting with yours. You are still trying to understand, but we are working to help with that.

The groups that you sent here, particularly the ones we have been talking about, must have had some experiences on their home planet before coming here.

They have visited other worlds and are acquainted with, to an extent, other primitive species.

I would think that their life path, because it is long, would involve moving about on this planet and perhaps other planets (over many years).

They have the option of leaving and returning from and to their duties here in the exchange program. They can often be exchanged for others.

In the initial exchange program, how many of the hybrids, after being here a while, chose to go back home?

There were over twenty beings in the original group, but many of them have been exchanged for others. However, the total amount has always remained about 20. Though they are not the original group, they can be because their life span is such. Sometimes they just tire of their assignment and wish to move to other stimulating projects, but they are also required to return again. It is like a refresher course.

We all have points in our activities, some enjoyable, some mundane, some not enjoyable, that trigger a need for change.

I sometimes get the impression that life on Earth can be boring. In using this term, it could be misconstrued as rude. It is not intended to be. It is a temporary assignment. It would be like being assigned to a small town and yet, occasionally, they must be transferred to other interesting places in the Universe and see new and wondrous things. Still, they enjoy their return. They are given that option. Like your humans on Zeta Reticuli, they are given options to travel, too.

Many choose to stay, and some travel, and they can return home or back to Zeta Reticuli.

It refreshes us to travel. As far as the concept of boredom is concerned, I have felt that much boredom is the result of one's own doing and thought processes. It would be up to me to find something to stimulate me.

Part of it is the telepathic community of 20 beings versus the normal interaction of a community of billions of entities. Often the stimulation and communication can be difficult to sustain beyond the stimulations of interacting with your planet.

That would be confining. I could see that.

It is part of what we have accepted. We have learned to focus within ourselves by using meditation to cope with that. We focus inwardly as well as outwardly in meeting our responsibilities to our mission on Earth. We have time to withdraw and focus inward to be with ourselves. It helps us to hold onto the essence of what we are while we are among humans, who often behave differently. This is part of the necessary contact. It is like being a missionary in your culture who is accustomed to the internet and automobiles, but must relinquish advanced technologies that individuals enjoy in order to live in a more primitive culture.

Thank you.

Contact

Is it possible for me to talk to Gen for just a moment?

Gen? Oh. He is not here today. Sorry.

That's okay.

We can call Gen.

It could be next time. I have a pretty good sense of you and Zestra, but my sense of Gen is not very strong. I know he has been contributing.

Gen has been the least known of the three of us. I understand. Maybe next time.

Whenever he chooses to come, I would just like to get to know him a little better.

He has been occupied.

We, Han and Zestra, I don't know if you can distinguish between us in our personalities.

I do sense some differences, yes.

We are the same, but different.

Yes. I think having a male and a female has made it more definitive.

It provides a broader view and a complete picture, which is appropriate and necessary. This is why I, Zestra, have been introduced into this relationship.

It is a lovely addition. Thank you.

Would it be all right with you if we expanded Steve's (Steve is in a state of deep hypnosis at this point) recall just a little bit regarding his childhood experiences? He certainly has some recall.

We can explore this, and see what is inside of Steve.

Thank you. He felt pretty certain that he saw inside a craft. Can you tell me how many times he has been inside of a craft?

The best he recollects is having no experience inside of a craft, but he has been near the exterior of a craft several times. He has been able to gaze in, but not permitted to enter. In one situation, it was urged that he not do this. He was attended by an entity who guided him and visited him on three or four occasions. He recalls this entity only once or twice.

Yes. He was fond of this entity.

At first, he was frightened, and the entities placed thoughts in his mind, and concealed their presence by fabricating a nightmare in his mind based on old mythology about humans. However, this was later found to be unnecessary as Steve's temperament was one that was open and accepting of us. As a result, the initial contact and the nightmares it created were traumatic for a time. Eventually, he reconciled those thoughts and memories and discarded them. The recent therapy with you clarified the intent and what actually occurred, letting him see that it was not intended to be an unpleasant experience. It was a welcoming one, introducing him to friends that he was connected with, although he was unaware of it. They were introducing themselves.

I am thinking that some of the children he played with may have been such 'introductions.' He had memories of playing with children, either across the street or next door.

He did play with children next door. There were many children. They were actually one family that had many offspring. This was new to him, so many in one home. Where he lived at that time, he was a lone child or with only one another sibling, who was an infant at that time. The memories of these children next door were used as a cover for a contact. It was with a craft where he was introduced to alien hybrid children. The neighbor's children were an effective way to conceal that memory. We often use implanted memories to cover the real activity. He was introduced to alien children and was quite comfortable with them in a playful way. He was accepting. In some ways, he was more accepting of them than he was with his first interaction with human children when he began to go to school. He does not understand why. He wanted to welcome them in the same way. Sometimes, he could be confused by the behavior of other human children. It was less so with the children of the entities. Maybe certain telepathic pathways were not available with the human children, and he felt a bit at a disadvantage or blinded from certain receptive thoughts. The behaviors and intentions of human children were not clear to him.

Was telepathy being used when he met the hybrid children?

Yes. This was denied him with the human children.

That would explain it.

He struggled with relearning social interaction without telepathy.

That makes sense. Thank you. (This also explains Steve's easy telepathic connection with the Zeta kin.)

How is it that Steve and I were so fortunate to contact the Zeta kin rather than the Verdants or some other species who may have ultimately been a problem for us?

To an extent, you selected the species to contact by having certain requirements to make the initial contact that was to be benign and not malevolent. That, in and of itself, made a selection which, though morally relative, we were similar in temperament and attitude and outlook. That is why we received your extended contact request. We facilitated this and intentionally locked out less fortunate species to create an exclusive contact with you. It was important that this was so, and for our pleasure as well.

Thank you so much for making that happen.

If this was selfishness on our part then so be it, to use a human value, but we wanted to be first in line, and that is what has happened.

I think it has served us well.

It has been mutually beneficial.

The Universe

May I ask about a planet called Iarda? This planet was written about by a person called Stefan Denaerde. *He talks about a planet being about ten light years from Earth. Is this contact a true contact, or is this a fictionalized account?*

Another human making a similar contact as us?

I believe so. He wrote a book. I think that beings on planet Iarda contacted him directly and told him a great deal about their culture, if this is a true account. Is this a true account?

It is undetermined. It is possible to be in contact with many species. There are many existing out there. Some we may not be aware of, to adequately answer that question.

They would be further away from us than your planet, I think.

We are approximately 35 of your light years from you in Zeta Reticuli as you refer to us.

Thank you.

Actually, they would be farther away from you than us. We are within a local group of stars that have stable nuclear fusion. The word is hard to.... The activity is stable. There are stable stars, which are conducive to life.

The fusion and the outer activity are in balance. Our own sun here, too.

Your sun would be considered in that category and in that group. You are very fortunate.

Yes. I enjoyed a presentation on the sun here, and it gave a good explanation of internal and external stability.

Other

I don't think I asked you about spontaneous combustion and human beings.

We have never discussed this physical phenomenon.

I know this happens. I do not know why. Does spontaneous combustion happen in other life forms around the Universe, particularly with beings who have a high-water content?

Spontaneous combustion in human form is caused by a heat source in close proximity to a larger volume of fat content in the human body. This can cause combustion when exposed over a long period of time. The degree that simultaneous combustion can occur with other species is incumbent to a degree upon their body fat and a heat source. In a physical sense it is possible, but the fuel for the combustion within many entities does not contain the same degree of fat for storing energy for survival. As a result, combustion is less likely to occur, but it is not impossible.

The ignition source, is that always an external source or is it sometimes internal to the body?

External.

Thank you for solving that mystery. You know of my friend Doc. He so much wants to participate in communication with off-planet beings. Still, I think he has probably finally accepted that there is other life in the Universe.

Life on other worlds?

Yes. I think he has finally begun to believe it is true.

If we had never met, this would be inconsequential to the expectation that life is everywhere, whether we met or not. Our contact validates that, but even if that was not the case, it is the correct and clear view that life is abundant everywhere, and you are one in a great chorus.

That touches me deeply.

The electronic noise that emanates from your planet is hard to ignore to the extent that light information can travel. In the last 100 years, we have heard your radio transmissions, due to our close proximity. We have heard the din of the many who, quite unaware, are heard beyond the surface of your world. This electronic noise has provided insights into your culture and explained many things that have puzzled us, as well as things we found to be enjoyable. This has been very beneficial in helping us to understand you through the artistic and non-artistic communication of your species.

A lot of that material broadcasted is really a caricature, usually exaggerated for the purpose of entertainment. If you judge the ordinary human from that kind of expression it might distort the impression of who we really are.

We understand and accept the difference between character and caricature and know that an artist will create exaggerated features, which are often for amusement in communicating notable characteristics. It is a human trait to sometimes exaggerate. We accept this. We often do this, too. Our communication here may create a caricature of our culture, but I, Zestra, as well as Han, Gen, and the others, have tried to give a well expressed and broader view of our culture. It is with pleasure and happiness that we do this sharing because I know it must be all wondrous to you. Understand that you are just as wondrous to us. Sometimes we do not understand how you can function without telepathic ability. It is interesting to observe the social dynamics, both individually and in mass populations, of how you conduct your selves. This has been the central theme of all our communications and sessions that we have conducted in sharing what we have in common and how we are different. In our differences, we attempt to understand each other.

Well, like you, I don't mind the differences. Differences make life a delight.

As it is with us. That is why, even if we may appear advanced, Doc should not be overly awed. I can understand his initial assessment. We find you just as interesting as you may find us.

(Doctor Lipson asked me, the therapist, to ask the Zetas if they could determine something important that he had been thinking about. They responded with Doc's memory of a patient dying.) You had plucked a memory from Doc's mind about him worrying about a patient and the patient dying. The question that I had framed was that Doc was looking for a specific event at a specific time. Of course, the death of a patient would be a primary event in his life. Doc, in his own interpretation of what is proof, was looking for confirmation that his big toe was giving him extreme pain. I had to smile, not at his pain, but that he was using that as an event.

We did not feel that was a significant event, though it must have been for his big toe. (Zeta wit.) We may have misinterpreted what he was seeking. We are not without our faults. Doc is a being of kindness, and one who deals with immediate concerns. Some concerns can be more immediate than a sore toe.

I met someone named Fred Thaheld who has been writing abstracts about quantum mechanics. I told him I would present a couple of his concepts to someone that I knew to gain a different perspective on his propositions.

This is theoretical?

Yes. Definitely theoretical. He poses a couple of questions:

  1. *Does or can consciousness collapse the wave form?*

In the context that consciousness can be immediately transmitted across great distances, which we use for communication across great distances, it is conceivably possible that a wave form can be reduced by that process. It would only be possible through a quantum means. It is the only physical reality that would occur. It is possible in that context.

His second question involves measurement and the human eye. I think he contends that quantum mechanics has a linear nature. I don't know if that is true, but he poses the question:

  2. *Does perception violate the linear nature of quantum mechanics?*

Quantum mechanics is not linear. It is the natural human perception in understanding it that it appears linear. I rephrase. Your human perception makes it appear linear. It is just your way of grasping it. However, quantum is not linear, but it is just a method of trying to understand it. I hope that has been expressed properly.

It has. Does perception interact with quantum mechanics? I guess you would have to say consciousness, really, which was addressed in the previous question.

Perception is a view of consciousness. No. the word is wrong. Perception is from consciousness, one viewpoint of reality. It is what exists, and our reality is only what we perceive it to be. The human mind is a mystery unto itself. It is conscious and aware, and it understands, as best it can. One of these understandings is how little it does understand, but it is right to do so. At that level, we are no different. The capacity of our minds may be different, perhaps more, but we still, in our awareness, only understand our Universe and the quantum aspects of it to the extent that our minds can conceive and comprehend it. Such is life. When I spoke earlier of the species that became pure energy, our great endeavor is to understand that one of the great differences is that we only *perceive* quantum mechanics. When a species becomes pure energy, they no longer perceive quantum mechanics. They become quantum mechanics. That is profound to us.

I think I suspected something like that, maybe not expressed as well as you expressed it.

If it was expressed well, it was not without struggle to express it, but I am pleased you understand now.

Perception is a physical term.

It is one that is divorced from the reality of what is, and it is difficult for a species to perceive it separate and apart from true reality. To become quantum mechanics itself opens up a whole order of magnitude to the reality.

That is perhaps different than when one returns to the spiritual form. My understanding is that they are able to perceive a great deal more, but that is different from being the process, the mechanics itself.

This is true.

(Han and Zestra continue down this line of query and share some of the questions that the Zetas have about these extraordinary advanced energy beings.)

We have other questions about these beings: Once they have achieved this quantum state of consciousness, do they become an absolute? Is it stable? Can they manipulate quantum mechanics for their own purposes, or do they just co-exist in that state?

Those are excellent questions. My humble question would be: We look at change as being part of a natural process of evolution. If evolution is still operating at that level, then one would think that change would be a part of that state of being. Do they continue to evolve?

(The Zetas continue to ponder larger questions.)

Would this intelligence be able to change the reality of quantum effects that, in turn, would affect the rest of the Universe? Would they be able to control that, and would it be for their own purpose? What would this represent for the rest of us? Perhaps they no longer have the need for a personal or species' agenda. Perhaps they have evolved beyond this. This appears to be the case. It is not affecting us. They are now merely on a new plane. You refer to change. We, within us, instinctually seek change, either across the existence of a species, or just within an individual day-to-day life. Change represents renewal and new beginnings. It is the driving force that helps us evolve and continues to do so. All species that are successful welcome change, at least on an instinctual level. Maybe on a cultural level, they do not always welcome it. The culture, being in the foreground of thought, can often crowd out the old instincts to move forward.

These are lofty questions and well worth considering.

Open Session for Comments or Questions

I believe it is time to bring Steve back. I look forward to the responses to your quantum questions.

(The Zetas share some passing thoughts.)

We are feeling rather somber in describing the human entities visiting here on Zeta Reticuli. (Reference is to Serpo Project voluntary exchanges.) I hope I have not left that impression, but it is somber for the human entities who are exposed to our many wondrous things. That is often what happens in such an exchange. It is accepted. We have self-discipline and also discipline within our mental community. With their community of 20 (the 20 humans living on Zeta Reticuli I) they relate to each other and understand their purpose and conduct with joy. On that note, I understand that you now wish to conclude, and Han and I will step back. Han wishes to send his regards as well. (Here, Zestra was the primary communicator.)

Namaste.

Namaste from both of us and to Gen in the future. Farewell for now.

Farewell.

Session Forty-One
May 6, 2012

Steve is sent to work with Scala, Lent, and Nicole, on a planet-scaping project in another dimension.

Han, you are cordially invited to speak with us today.

Hello, Mary. This is Han.

Have you been well?

Yes. Gen is here. There are just the two of us today.

Delighted to have you, Gen.

Hello, Mary. This is Gen. Nice to speak with you. Normally, I am quiet and, in the background, but my thoughts contribute to the others as it is in our society. I am here with Han and available for any questions that you might have. I will try to answer them.

Thank you. I appreciate that. There is an old saying that if one is talking, one is not learning. I am thinking that you have been learning a great deal, but also sharing.

I am relatively inexperienced compared to Zestra and Han. By permitting my presence here, they feel that I am worthy to share in this experience, so that counts for something. I am observing and wondering about all that is occurring and enjoying the interaction between us, between your kin and our kin. It is across unimaginable distances that we can share like this. The physics might seem almost incomprehensible. Actually, it is a natural and special part of the quantum physics of the Universe that permits this. It is not miraculous, but it is miraculous at the same time. It is instantaneous across great distances.

Does this mean that the communication is occurring, in a way, outside of time and space?

It is not constrained by the normal concepts of time, particularly in a linear way. Time has depth. It is like a wall of water in which you may see time from only the surface vantage point, but it has great depth underneath it and has many layers that move as one. This lends itself to our conversation, in the curious nature of quantum physics, and enables the link where we can both speak across the distance at the same time, without any lag in communication.

How does it feel for you to be communicating through Steve's vehicle, the vehicle of his body?

I feel Steve has provided the most spacious room where he can graciously provide the hospitality of his mind, and yet it is, in a way, still very confining. I know that is only due to him stretching to accommodate us to make us feel comfortable. We could communicate in many layers. Han and Zestra have spoken of this. We have, in your communication over the past weeks, learned to flow with this. It is not limited to just voice communication. Many thoughts are given to Steve, both thoughts that are related and the many thoughts that are not related. He remembers and describes them to you after the sessions are over. Steve is very detailed, and that has been very helpful and unprecedented. It fills in a great mosaic painting, a tapestry of the paradigm of our reality. It is not that different from yours.

We have the same physics, rearranged in different patterns, but the same physical patterns. Our mountains and rivers are the same. Our sky may look different with different constellations, different colors of sunsets and sunrises, new stars, but stars that are very remarkably close in their fusion stability. Over a long term, this is important for the development of life on the planets that surround them. Han is agreeing with me. I am enjoying this. This is new to me.

Gen, have you ever had an opportunity of sending out a thought form, causing a slight electrical physical reaction between where you are and where I am? I don't know if that was very clear. Sometimes Han and maybe Zestra will do this.

A charge?

Yes. An electrical charge. Sometimes I feel it in my hair and scalp when they stop by for a moment. It is just a way of saying, 'Hello.' Have you ever had an opportunity to do something like that?

I confess that it has been the mischief of Zestra and Han. They are not necessarily in any one place at one time. It is as if one entity is almost omnipresent in their thoughts, and some of these phases manifest themselves in an electrical field. It is with the thoughts of regard and happiness that they will often just give a gentle reminder of our presence, among of course, many other spirits who also frequent your home and your soul. The proximity to these other spirits, as they come and go in their visits, is like neighbors stopping to say, 'Hello.' We are doing the same. I, too, will join in that chorus so I can share in the mischief, if you wish to permit this. They say I can do this.

(Therapist laughs.) Yes. I think Steve would enjoy it, and I would enjoy it. I would be happy to acknowledge you, also.

I am more timid than the others, but I am changing fast.

I am quite comfortable with acknowledging entities, and you would be quite welcome.

Thank you. We will continue our visits passing by and through. We will admire your roses in your garden and your spiritual aura, which is a pleasant landmark along the road of the Universe. In that poetic way, we will show that the three of us have our separate, independent thoughts, yet blend in the same consensus of thought.

I am so pleased that you have joined us and stepped forward.

As you are aware of our place in this galaxy, I was wondering if you would tell us about any sentient beings who are close to this planet. I assume that life exists. I assume they are aware of us, and we are not aware of them. Perhaps they might be in the supercluster of what we call Virgo, which contains what we call M33 in the Milky Way. If you know of any non-human sentient beings in our galaxy, I would certainly like to hear about them.

There are other sentient life forms this galaxy in which we co-exist. It is just one galaxy of countless others, but we move together in a cluster group of other galaxies. Within our galaxy, life abounds in many forms, but, predominantly, the life forms that assume a humanoid appearance, to use your term, are the ones, which are necessary to create technology to begin the next threshold of evolution of travel beyond your home planet. We have this technology, and you have begun this process. Many other countless civilizations have done the same. They have coalesced into many areas of influence within this galaxy where, in a peaceful way, there is a dominant form of entity.

Your area coexists in an area that we, for the most part, control and influence. Control is not the correct term. It is just an area where we coexist. You are not aware of the 'control,' which is mainly to maintain the status quo with other species to provide a safe harbor with the species that coexist and communicate freely together. We are always vigilant for species that can form renegade groups, which, as you know, have abducted humans illegally. This runs against our ethical values. We have commented on this before. Han and I are speaking in a chorus now (Zestra, Han and Gen), and this has been discussed in the past. This is one example of why we would use the word 'control.'

It is not a 'control' of the human species on their planet so much as providing an umbrella to permit growth and the commerce upon your planet to exist and continue unfettered. This is primarily why we moved quickly to establish contact with you before others could. It was also that the criteria you provided in who you wished to meet was very compatible with our mutual needs. You wished to make contact, and we wished to make contact as well.

Spirituality

There is a state of oneness that you are able to achieve with your planet. I can sense that oneness when I am in a forest. I haven't thought so much about being in harmony with the whole world, but I know the vibration of the planet exists, and that there is probably a consciousness. What are your thoughts?

There is a collective consciousness of everyone, and yet there is still the individual. We have achieved, in a spiritual and political way, what you have yet to achieve. Do not despair. You may not have, as some of you on your planet have feared, a one government for everyone. In your species, there is not so much a spiritual unity with yourselves and the Universe. Someday you will see how much you all have in common in your basic elements (photons, electrons, etc.). We are not different from you in that respect. This fundamental perspective can be easily overlooked. This is one that, as your perspective grows, may take root in time. Humans do not normally view life from the level of photons.

Contact

You are aware of our friend George. He has started sharing more now and gives me lots of food for thought.

Yes. He is our friend, too.

He has his own perspective, and one of his larger concerns is that those who are supposedly in charge of each species may have agendas that they are hiding from their general populace. I don't know, in a telepathic community, if one needs to be concerned about that potential problem. I don't know that the leaders could hide anything. I have to leave him to find his own way through that process.

The telepathic community of minds often has only a distinct range of influence. Beyond this, other views can form, and often they coincide. Often, they do not. Also, various species have formed, in their own perspective, a reality of their own needs for material and technological superiority.

They are telepathic, are they not?

Yes. There is the limitation of distances, that many, with quantum theories have mastered, as we have. Also, some species are more telepathic than others, and some can be deceived. It is possible to collectively deceive another species, but this, in time, usually evaporates. It is usually a temporary illusion that can be created, if one was to use blocking, a human term, but it is more complex than that. For the most part, each species is dealing with their own needs of supplies and resources and wishes to control resources for their own security. We are talking in quantities that you cannot yet imagine. As a result, even though communication is free, there are still interspecies' desires and competition.

Are they able to hide things from each other, within a species?

It is possible to create an illusion. We have often done this with humans, but humans do not have the benefit of comparison with other impressions from the same species to coalesce and form an understanding. It is possible to create illusions. It is not in our nature to do this within our species. We know that some species are clever, and it would be naive to not be prepared for this possibility as well.

I have encountered them (false memories, blockages) with my clients.

Yes. I am sure you have.

The Universe

I was wondering if either of you (Han or Gen) have ever come across some sentient beings who told you that they existed around and before the beginning of this Universe?

There are beings who, in actuality, may have. There are other boastful species who claim they have, and yet they have not. It is merely their pride in saying that they even transcend the Universe. It is their way of impressing others. Yet there is a new species that we have encountered, which Zestra and Han have commented about, that we again have referred to as the Spiritual Beings where they have almost become the Universe itself, returning to the basic elements of mass and energy. In this form, energy could transcend universes, from one universe to the next, existing before the creation of this Universe. Perhaps this is the way it was before universes came into being.

I like that idea. Really, it makes sense.

I am explaining this in a way that you can understand, using your language. Your language is not limiting. It is merely condensing our thoughts into a few chosen words instead of the normal stream, but I am growing accustomed to this. In its confining way, it challenges us, which is intellectually stimulating.

It is an attribute of poetry to take a great deal of meaning and convey it in a few well-chosen phrases.

Yes. That is a beautiful analogy. Our communication is poetry in that sense.

That is a lovely thought. I have noted that from time-to-time. You have come to know the moral attributes of Steve and myself and to know that our intentions come from kindness and curiosity as well as the pleasure of stretching our intellect and satisfying our curiosity.

This is true. These values are not inconsistent with us as well. This is perhaps why we are so forthcoming in sharing information with you.

You know that humans, along with many other life forms, have a need to play. Would you give me an example or two, aside from mischievous electrical contact, which I enjoy, of how you might play when you are with your Zeta kin?

Often, the more complex the mind, the greater need for the simplicity of play. In our realities, we can create other places to rest our thoughts and our concerns, thereby creating a therapeutic diversion of our consciousness. You use the term 'vacation' or 'holiday.' You will often expend great energies on these vacations. Sometimes it is puzzling that you do not rest on these vacations. In time, we began to understand that it is merely changing the focus of your thought patterns to another focus. In that respect, traveling to another physical place is often not necessary, but it does help in that there is more to the senses than we experience in a physical sense. Our conscious minds can be permitted to rest, and these other areas of the senses, in other locations, can be enjoyed without having to create an artificial menagerie of a separate reality. We have other places where we can go to relax. We even visit your planet as vacationers to explore and observe, although many visits to your planet are of a scientific nature. These visits include maintaining a presence for other species to see who comes here. Other planets are also available as places to relax.

I see, in my mind, other species sailing out on lakes, in a very ancient way, as a form of relaxing or maybe just floating on the water. I know you are not great swimmers, but is that a form of play for you?

Yes. We do not have the buoyancy that your bodies carry in liquid water, but we do have apparatus we carry with us, which can make us float artificially, regardless of whether we are in water or not. We use old-fashioned wind propulsion. Often it is not, compared to other means of transportation, quite how you see sailing. It, has taken on a new meaning, where it is not the fastest form of travel, but it is certainly one of the most pleasurable. It helps us in our natural mind set of being one with the planet. We can feel the boat under us, and the wind and water and become one with it, using it to the best advantage. In doing so, we form a harmony with all these elements. It is not different from space at all. It is the way of our species coexisting with our surroundings. It is the fundamental philosophy of our species that we are one with the planet, not at odds with it. We are a part of it. Many of your Native Americans and Aborigines have known this, and, as has been mentioned by Han, our philosophy is very similar to their philosophies. Sailing is one form of non-effort or recreation. We also learn to levitate, and sometimes we can transport our minds to other places in consciousness. For those who are not able to physically travel, there are many realities and diversions available to us.

Have you ever tried something like paragliding where you use an artificial attachment to your body?

I was reading your thoughts. I thought of paragliding.

But I don't know if you use it or if you use something similar.

We used to do this, but we were having too many injuries. However, we have found other means with which we can drift with the winds, without the need or dependence on aerodynamic forms. We can levitate. This is quite natural to us now. Mostly, in our explorations of your planet, there is a lack of physical trace of our presence. This is partly due to our ability to float without having to touch and leave marks on the ground.

I have seen a videotape of what I believe to be a Zeta kin doing that. He was floating across a front yard.

We will often float and will seem like spirits in your culture. It is actually our means of moving freely. It is partly recreation, but it also forms a serious purpose in not leaving trace evidence behind. Many of your UFO organizations stress finding physical evidence, and we are very careful not to do this.

Because the Zeta kin originated from birds, I was thinking that paragliding or riding the thermals would be a very natural, familiar and pleasurable thing to do.

The winds still exist, and we feel it on our bodies and on our faces when we move through the atmosphere. We acknowledge our heritage and evolution in this area. Being in the sky is part of what we are as beings.

We have something that we call an Ort cloud on the edge of our solar system. What can you tell me about it?

It is a nebula, a nursery for new stars. It is also a nursery for dark matter, as you refer to it, that constitutes most of the physical Universe.

I hadn't thought about that. a nursery for dark matter.

This is our understanding. It is one of many places.

You have corrected an error in my thinking. I thought that dark matter simply always existed. You are indicating that it has a birthing process.

It has matter as do other forms of matter. It is created from something else. It does not come from nothing, even though it may seem to be the case. I am speaking in very general terms now, perhaps too general, for I know your intellect could probably comprehend more. I (Gen) am sorry.

I was wondering if you and Han would like to take us on a little journey to another kind of reality, perhaps another dimension, just some kind of journey that you think might be interesting to us.

Another dimension might be possible. We could choose one of multiple dimensions. Steve is hesitant, but I am pushing that back. He is not confined. We have visited many other planets that might seem mundane. I will attempt to take us to another dimension, one that permits us to communicate the way we have across distance at the same time, where everywhere is the same time and every time is the same place. It is difficult to express in your minds how this feels. There is a sense of great freedom where one is released from the bonds of physical dimensions and time. This is where quantum mechanics rules, where rules of old do not apply, and old concepts of reality are confounded. It is a place that you do not travel to because it is everywhere and nowhere at the same time. Forgive me for couching my comments in riddles.

'A state of being' may be what you are intending to express.

Yes. It is a state of being. Thank you. It is one of bewildering vastness, even greater than the Universe, and within this void exists the materialities that we call universes. There are as many universes as there are galaxies. It is difficult to comprehend, as Steve is bewildered at trying to comprehend, but he is not to worry. He is trying to move the water of a fire hose through a garden hose while he is holding tightly to the hose so that the pressure does not overwhelm him. He is not under pressure or strain or difficulty. It is just new.

(Addressed to Steve.) Just allow it.

I am focusing that 'state of being' now and trying to encompass all of us present to try to help comprehend this. In that meditation, I must release and come back. It is a strain to do this, but everything is fine. It was a short journey, but we did not have far to go.

(Therapist laughs.) I will call this the 'State of Endless Potential.' Perhaps, during my meditations, this can be expanded a little better, but at least I have a sense of it.

I think you can easily accomplish this.

I sensed a single tone during this process, but it was a tone like no other. It really spread everywhere. (Therapist meditates and receives a strong sense of a magnetized void with perceptions at the subatomic particle level, lots of movement, and endless potential for developing material and nonmaterial life forms and non-living material forms - a state of potential in its most elemental form.

I am not so much taking you to another place as much as I am relaying that which you see, we also see. It is still a mystery.

The fact that there are other dimensions is quite acceptable to me. I have probably journeyed to some of them, particularly in dreams. This one is new to me.

Our physics teachers are very comfortable with these concepts. It is the basis of much of what we understand. But traveling there is often another thing.

Ah, but what fun. I didn't know that Steve was uncomfortable and felt he had to struggle with it.

He is not struggling. He is just daunted by the vastness, like standing on a precipice of a vast canyon.

I don't expect to always see a horizon. I sometimes forget that others need to see a horizon.

It is a reference point, a reference in reality, as well as in time and space.

It is interesting how human beings will often become nauseated if they lose their sense of the horizon (car sickness, air sickness) when they are traveling.

It is their reference from where they were once to where they are going. It is an emotional reference as well as an intellectual one.

I remember a story about a pilot, Bruce Gernon, who, in 1970, left Andros Island in the Bahamas to fly to Florida. Somehow, after flying through an 'electronic fog,' he arrived in Bermuda, way ahead of schedule. This was impossibly fast for his airplane. I don't necessary think that he and his crew had been abducted. I think they may have passed through some kind of rip in time or some anomaly. Do you have any ideas about how they would move so quickly through time?

There are certain rifts in dimensional time and space. It is possible that they passed through one of these. For them, time passed normally around them. They had gained or lost time. I sense from your conversation that they had gained time, and they basically arrived before they were due. We have observed species that have abducted humans. To demonstrate for the skeptics, one species has often left puzzles behind, i.e., one woman could be abducted in one part of the continent and returned at the other end of the continent by a means that, in physical human terms would be impossible. That would be the puzzle left behind for those intelligent enough to recognize it. Still, when the human mind grasps the situation, their dogma will push away their grasp of the situation, and they will surrender, once again, to the possibilities of other things.

So many do grasp their dogma in unusual situations.

We admit, in our dealings with Earthlings, that we have used this tendency of humans. By distorting another species' sense of reality, even if they should see us, humans will return to their dogma for explanations, and we are explained away as something else.

May I ask you about a person named Annabelle (pseudonym) who is one of Steve's acquaintances?

It is possible.

She seems to get a little upset when Steve talks about working with abductees and trying to help them. She will leave the room when she can. It made me wonder if she had experienced something herself, or perhaps she was afraid that Steve was delusional.

Annabelle is a very hard-working career person. She has, in her own personal life, dealt with depression and finding her place in herself. It has been a long, personal struggle for her. With Steve's good intentions and concern for abductees, as you are also, she becomes concerned that somehow Steve might follow in her footsteps, living a life with fear. She fears for him due to her own struggles with depression and her sense of identity. In her reality, she needs all of her ducks in a row. Everything is ordered and correct. Even when her home is messy, in her mind, she finds comfort and security in the sun rising and setting at predictable times. Such anomalies as abductions, can threaten her sense of order and predictability. She may find it threatening to her paradigm of reality.

Thank you. I understand that and also that some people need to lead their lives in a very controlled, orderly manner without much room for other things.

She has worked hard to establish that. It is the pattern that Steve has also embraced, although it has been challenged in recent months. These have been positive times in stretching what are his expectations of life. The reality of alien visitation is one that is too inconvenient for Annabelle, and she is not alone in this, as she represents the bulk of humanity in that respect. It would be too much change too quickly for some. Yet it is occurring. This, for example, is why we often fly off the ground so we do not leave footprints. This all ties in together. We are very careful not to disclose our existence to others who are not yet prepared for us.

Yes. I have been aware that you respect people's paradigms.

It is a respect for their paradigms, but it is primarily to permit us to achieve our goals more effectively. If they do not know of our presence, it makes our lives easier in accomplishing our tasks.

You have already partially answered one of the questions I was holding back. The question was: What are some of the things that you hold sacred or of paramount importance? I think I know some of them. Are there any that you would care to share?

- To promote life and happiness
- To exist at one with the Universe
- To cohabitate with other species in harmony
- To pursue the unknown with all of our energy, thought and reverence

It is with knowledge that those understandings make our coexistence with the Universe more intimate. (This desire to know the Universe with reverence seems to be an extension of the reverence that the Zetas hold for their planets).

Open Session for Comments or Questions

I have just one last question about your home planet and the sounds that your home planet emits. I assume there are magnetic loops, but are there other less obvious sounds that are emitted by your planet that might be based on a kind of consciousness?

The physical resonance of a planet, when it is a large enough physical body, can act as a collector of consciousness energy. It is difficult to define this. It acts as a depository of energy. Energy is everywhere. What is not matter is energy. The fields in consciousness can coalesce in these regions due to the conscious intelligence that exists and thrives there. It acts like, in a sense, a parabolic mirror, reflecting and receiving energy. It is a form that can be exploited and used for greater consciousness and understanding. Often our species has tapped into this. In a way, it is an amplifier of consciousness, and it has helped our species as it has probably helped other species as well. This understanding, which you have begun to discover, is the same on your world as well.

Your planet has a great deal of water. Is this part of the process?

The water has its own special dynamics, but predominantly, the larger mass of our planet, though it has the same land mass, has a larger surface of water. Beneath our crust we still have tidal and magnetic forces consistent with other cooling planets. We exist on the crusts of these planets, as do you on your planet. I am sure you can relate to this. They form dynamic engines for many properties, magnetic, electromagnetic, radiation, and even an engine for consciousness.

I really enjoyed your talking about this and will probably bring it up again.

Please do. All questions are welcome.

Steve and I look forward to any energetic visits.

Yes, we shall continue this. We will also provide the answers to your questions, if we are able. There are many answers we have not yet solved ourselves.

Ah, but the questions are so enjoyable. With your kind permission then, it is time bring our session to a close.

Gen: I look to Han. Han's thoughts tell me that it is time. I understand and thank you for the gentle time to exist with you in fellowship.

Han, thank you for bringing Gen to this session.

This is Han. Gen is a young one. We are well pleased.

He has a lovely energy. Thank you. With that I wish you a fond Namaste.

Namaste to you from both Gen and myself. Farewell until our electrical energy plays with your hair once more.

(This last curious statement from the Zetas is a reference to their way of making their presence known. After the first session Steve and I had with them in 2011, they continue to occasionally cause a small electrical whisper in my hair near the crown. It continues to happen about every week or so, just letting me know that they are aware and keeping contact.)

Session Forty-Two
May 20, 2012

(Using a hypnotic suggestion, this therapist sends Steve to work with Scala, Lent, Nicole and others on a planet-scaping project in another dimension. This is his terra-forming project on planet Bonestall, as Steve has named it. This is the completion date for this project. This project only exists in the mind and is a method for removing Steve from the exchange between the therapist and the Zetas.)

Han is cordially invited to speak with us today.

Hello, Mary. This is Han.

Hello. Thank you so much for coming to visit us today.

I have been anxious to meet with you. At least that is the human term. It is not really anxiousness. It is really just the anticipation. It is good to be here again.

You are the only one in attendance today?

Yes. The others are occupied, but they are within telepathic range. If you wish them to join us, they would be happy to do so. Know that Zestra is now approaching.

This is Zestra. Han was not excluding me, but I just stepped in, and now there are the two of us here. Gen is occupied.

Han is speaking now.

Earth History

When one becomes a President of the United States, the new President undergoes a briefing about the State of the Nation. It is a secret briefing. I understand that President Carter, after the briefing, sat at his desk and wept. I think there are five Presidents still living. What is it that these Presidents found out that is so overwhelming?

It is a combination of secrets unmasked, and the clearer picture that forms is often a profound one. What was previously known had many complexities as to the true nature of things. This crying, combined with the burden they assumed, is often a release of stress at both the new knowledge and their own responsibilities of office. It is a rare and unique experience. Few would empathize, having never experienced it. It is an experience of only a handful of human beings. Their expectations and knowledge are stretched, and they are enlightened. They become aware of many new things and receive confirmation of some old ones, often enhanced with more detail of relationships between countries, treaties, and potential for danger that had been kept in check. They receive secret information and communication access that is unprecedented for them. They know they are at the center of this, and now everyone is looking to them. They often sit in mediation and pray that they make the right decisions. They have strived hard for this privilege, and now that they have arrived, they have a different view from the top of the mountain.

President Truman said that he felt like he had been hit in the head by a bale of hay when Roosevelt, the then President, died in office, though not unexpectedly. President Roosevelt was effective in the War as surely as any soldier in the field that he commanded. He commanded with great strength, and he was held in profound respect by his fellow humans for leading them through very difficult economic times and, in matters of world security, eradicating the 'great evil' as you interpret it. It is one that we also recognize as an evil, if our values are to be good and noble. The ugliness of much of the War is too profound to comprehend. I want to push the thought of millions out of my mind. I am overtaken. I wish to change the subject. You understand. (Han is overtaken with emotion. Zetas do have emotion, but usually minimize them.)

I certainly understand. To change the direction of the subject, one of the concerns I had about Jimmy Carter was that his initial briefing may have shaken his religious core.

The briefing concerning the visits of my species to your Earth?

That may have had that effect.

I have found that it has not shaken my spiritual beliefs, but it may have shaken his spiritual beliefs.

It had two effects: It compelled him to see that the old ways were obsolete. The new information helped him to focus and clarify. In this process, one prioritizes beliefs and knowledge into what to disregard and what to hold onto. It is a thought purification process. It helps to define what is the essence, and what is merely rhetoric or silly concepts. The result is that the knowledge of life existing on other worlds truly helped him to realize that neither his beliefs, nor his value as an individual was diminished. Rather, his beliefs and value are only magnified.

Well, that is what it did for me. I certainly hoped that it did that for him, too.

This is correct. This new knowledge magnified some of his beliefs and, with the clarification, it told him what to disregard in his religious beliefs.

He has continued his life, with his charitable heart and charitable actions.

Many of his decisions were controversial, and some were highly criticized. This was a new administration, compared to Reagan, who was a strong contrast. Carter provided a time for rest and healing after earlier administrations. This helped to refocus the nation. However, Reagan's role was just as important when it was time to again pick up the responsibilities and to end the cold war. It is a remarkable time in human history that you and Steve have experienced. It was one of fear and of hope. It took an actor to be able to empathize with the human condition. He used his skills as an artist (actor/communicator) and, with these skills, he expressed, with great communication precision, the delicate issues at hand, many of which were successfully resolved.

President Carter continued in his role as a facilitator and ambassador. Many of his energy policies were not efficient for the times. These policies have had a negative effect. One such change was his nuclear policy and his role in a nuclear Navy under Rickover. Carter felt Rickover was capable. One great mistake was made in 1977 when President Jimmy Carter signed an executive order that banned the reprocessing of nuclear fuel in the United States. (The rationale was that the Plutonium could possibly be stolen,, and terrorists might be able to use it to make atomic bombs.) France did not follow that policy. France created breeder reactors for reenergizing nuclear fuel for use in nuclear power plants. Frances uses this cycle. It is important and, as a result, they have a simple, functioning system that provides significant energy and, politically, does not make them dependent on foreign countries to the same extent as the United States. Instead, the United States stores its fuel, and much controversy occurs as to where to safely store it. Had they followed France's example and taken the fuel and used it in breeder reactors, they would have created an almost perfect dual system of limitless energy. The United States has failed in this policy if that was their goal.

Perhaps we can still reconsider and use France's type of fuel regeneration.

New designs of reactors will come to the fore. However, there is still an irrational fear of nuclear energy. We have monitored your nuclear plants and your weapon systems. Considering the primitive designs, you have been good stewards and responsible for their safe operation and storage. Other countries have done the same. Some have done less so. The most efficient and friendly for the Earth's environment is nuclear energy. It will provide almost limitless power and permit land to be used for other purposes such as agriculture, population growth, and for natural spaces.

What do you mean by natural spaces?

To preserve the natural environment. In many wilderness areas solar energy is very appealing, yet it is a very inefficient form of collecting energy. It requires a great area of space that could be devoted to other purposes, such as agriculture.

I wasn't sure about the production of solar panels, whether their production may be unduly toxic to the environment.

The chemicals involve in the development of solar energy can cause more harm than good. The development of the technology and the damage that it could cause, is a poor second because it is not as efficient as nuclear energy. Ironically, nuclear energy, in the public mind, is seen with irrational fear. That is very understandable. You often confuse nuclear weapons with the more peaceful uses of energy.

I am glad you brought that subject up because you are correct. Humans do tend to generalize, often inappropriately,

I would like to talk to you about a "catch 22" situation.

Ah. Irony.

(Therapist laughs.) We humans seem to be suffering from a vitamin D deficiency, particularly here in North America. We were told to slather ourselves with sunscreen to prevent cancer. Here we are, needing sunlight for the manufacture of vitamin D, being told to not let the sun touch our skin excessively. That is our 'catch 22.'

I understand. Your society has changed. Many now dwell indoors, and their immunity to sun exposure has been diminished. Your species has been slow to adapt. You are lacking vitamin D provided by the sun due changes in your society patterns. At one time, many of you worked outdoors and were exposed to more sun. You built up a tolerance, which also was very sympathetic toward acquiring sunlight to increase melanin and vitamin D. The shift to indoor living created an awkward 'catch 22,' as you have indicated. Now that you have lost some of your tolerance to the sun, it exposes you to a higher risk of skin cancer. The solution would be graduated steps to return to an outdoor lifestyle. However, this does not seem likely now.

You are correct. Those who live the indoor lifestyle often become uncomfortable in extended exposure to sunlight.

At the same time, now you are more required to be sheltered from the sun (work settings). The energy from the sun has not changed. It does not have rays which are any more or less harmful than they were 100 years ago. This is also combined with new knowledge and new perceptions of reality. You now have almost too much information. For thousands of years your species existed well with the lack of knowledge about the sun and its rays, though you did know what caused sunburns.

Do you require vitamin D and exposure to starlight?

Yes. To a degree.

Your skin has now adapted so you don't have our issues?

Our skin absorbs more, even though our skin is light. Darker skin absorbs light more efficiently. We have chemically adapted to compensate. Our skin also acts as a supplemental form of respiration. We have talked of this in the past. Even though our lung capacity is small, our skin is able to respirate as well, to a degree. This is why we are not very good swimmers. It would be dangerous, yet we have means to adjust for this.

You have helped me to understand a number of things. Thank you.

There is human syndrome called Acquired Prodigious Savantism, in other words, those who become savants later in life, rather than at birth. In some, depending upon brain damage or other causes, certain parts of their brains are activated, usually to the detriment of other parts of their brains, and they may become gifted sculptors, mathematics, musicians, etc. This may happen without any more exposure or training other than seeing or hearing something one time. Does this indicate that there is a great deal more stored in our genetic structure that can come pouring through, fully developed, that we do not know about? Perhaps we are putting too much emphasis on education.

There is a natural preoccupation with human origins and the future of human development because that is the only paradigm you know. It will adapt in its own course and follow the natural progression that is often seen in evolution. Does this tend to answer your question?

No. It is just surprising that someone through simple observation can take that directly into creating incredible replicas in the finest detail. It is as though the requirement to learn all of the 'laws' of art is not necessary for certain individuals if the brain takes a different route. I think there is more to learning and what is available through different means than I ever imagined. (This needs to be explored more fully.)

Zeta Reticuli History

I would like to know about the experiences of your first colonizers as they moved from their home planet to their first permanent colonization of another planet, including the driving factors to colonize off-planet.

We have to go back to what is almost prehistory for us. This was a time before we had strong population control. We began to migrate to other nearby planets with more living space. The challenges were hard. This was understood and expected, and the technology then was more primitive. The challenges were harder. Now we have the skills and the energies to make that process much faster and easier. We only moved to uninhabited planets. Invading an inhabited planet was against our moral values. We have never invaded another planet by force. This is a Hollywood concept. It has occurred with other species, but it is rare. There are actually easier ways to eliminate a population, like tossing an asteroid into a planet's atmosphere, or even destabilizing the nuclear fusion process of their sun. However, this is a gross, immoral use of power. While such power exists, and it is not difficult in our technology to do this, we would not. We have other focuses.

In our early migrations, we migrated to nearby planets that also orbited Zeta Reticuli I and II. The distances between the two stars are vast, but relatively close in the greater conceptions of distance. Often civilizations will be blessed with nearby astronomical bodies and will cross the 'river' of space with evenly-spaced stepping stones (astronomical bodies) to enable easy transport from one planet to another. We were fortunate in having that. We could make a logical progression from one planet to another. Terra-forming was very difficult in the beginning. We were expanding to a new order of magnitude from a planet where the entire energy of the planet was focused and in harmony. Now we were expanding to a new capacity of managing the energies of an entire solar system. This is a daunting task and one that does not occur overnight.

You already knew how to fold time at this juncture of colonization?

No. We did have the ability to travel great distances quickly, but the distances were not so great, compared to now. The new age, when we folded time, released us to almost unlimited speed and distance, where distance basically disappears. The farthest star was as close as one hand to the other in quantum space. Earth sciences think only in linear time because that is the extent of their experience so far. This is both correct and incorrect at the same time. There are great distances. This is without doubt. The same physics applies everywhere, yet time and space can be distorted to cheat those distances.

In the beginning of your colonization efforts, did you only go to planets that had sufficient water, atmosphere and other resources?

Some planets were selected near us that had these essential, mission-critical elements, which were needed for us to sustain ourselves. However, some needed to be enhanced with increases in water and atmosphere in order to create and sustain an environment suitable for our species. Each planet would naturally be unique and not like our home planet. Our biology reflected our home planet's environment. With the new planetary colonies, we needed to change the environment to suit us. Care was taken so that local life would not be displaced. We took great care with this. We had to genetically adapt our biology to create systems and metabolism that would be adaptable to a wider variety of environments.

As you made these genetic changes to adapt to new planets, did the Zeta kin on the home planet still see you as kin?

They were not mentally different from us though there were differences in our physical attributes. We had well-established telepathic abilities already. We were all one. The physical differences were slight, accepted, and understood, but our minds were all connected. That is what defines our species. We have great tolerance for diversity. We are not that diverse.

My original concern was the separation from your home world and the resulting loneliness as you moved out onto other worlds. I see now that it would not be a concern due to your telepathy.

The connections were mainly local, within the communities on each planet. The distances between planets were too great for telepathic communication at that time. Our abilities were not that good, yet. We had not evolved to a state where we have our masters who can focus those thoughts in quantum space to interconnect whole systems. This is a special skill. It is unique. It is an ability that is not for all in our population. Some have acquired this, and they use great skill and discipline to do this, much like your monks who focus and meditate. The monks are star travelers in mind, as well earth travelers. They also act as receptors and transmitters of communication, along with technology.

This gives me an even greater appreciation for them. Thank you.

This is their role in our society, much like I am an ambassador, and Zestra is a healer and facilitator. We have, as we have discussed in the past, many roles, but there is a high mentality, as you refer to it, yet there is also individuality. It is like one plays the strings, another plays percussion, another plays the brass, and together, the many professions form many layers of beautiful music, creating one song that expresses the will and thoughts of the artist that created it.

Contact

A few nights ago, I received a call from a friend, Dr. Lipson (Doc). He wanted to tell me that he found a scoop mark on his thigh. It was not there when he went to sleep. Doc is well aware of what constitutes unusual markings on a body, and it was nothing he had ever seen before. Did this scoop mark have any significance?

He was visited. It was a brief visit. He has been touched before. He has often wondered, in bewilderment, why he seems preoccupied with this topic. There are echoes from the subconscious, memories unconsciously recalled, that compel humans to be drawn toward this interest. Doc has been a facilitator. We will often read his thoughts. As a human, he is a highly intelligent human being. This is, combined with a great compassion and desire for healing, that Zestra finds most attractive in him. Zestra is a beautiful being. She is willowy and, though it may seem strange to you, she is very attractive as a spirit of healing and service to her Zeta kin.

Doc was visited briefly and a sample was taken. Time was folded so that the element of missing time was eliminated. Just before he was taken, he and his wife were deliberately planted with a thought to sleep in two different areas of the house so his wife could sleep soundly. He was taken, but not in the normal abduction. He was taken by us. We do not perform against people's wills, nor are we very invasive. Our skills are such that most of what we wish to learn comes from the memories and thoughts. These were downloaded without him knowing it, and he was quickly returned. Time was shifted back. The sheets and blankets were pulled up snuggly to create normalcy, so he never knew he was temporarily missing.

Was he gone more than an hour of our time?

An hour only. He was not required to have a full physical examination, but he did have a brief checkup, as he refers to it. A sample was taken, and his mark rapidly healed. However, him being a physician, unlike most, he noticed it, and he also noticed how fast it healed. Many do not realize that, but that can't be helped. Still, he only recalls questions, with no answers.

The first time that he was visited, was he a very young person?

I am looking to Zestra. We are accessing it. Yes, he was taken as a young boy from a cosmopolitan area, maybe Pennsylvania, from an industrial area. Perhaps it was New York. I am not sure, but he was young. He was dissatisfied with many traditions and precepts. He could see beyond them, due to his intelligence.

Yes. I think he had a Jewish background with lots of traditions.

Yes, he did. He respects the Jewish religion, but he felt very confined by its patterns of religious expression. It did not seem practical to him. He is probably correct. Yet rebelling against this helped him move more clearly into what he wished. We contacted him as a boy, so he is special. Regardless of this, most of what he is in character and in mental capacity, is all to his own credit. His genetic heritage determined what he is. We do not take credit for assisting him. He arrived where he is on his own. Nonetheless, we were there with him. We were and are aware of him, and we hope that will give him solace in understanding why his views are what they are. I hope I am understanding your inquiry and expressing my response correctly.

Yes. Because of his mind, his ability to see beyond, was he able to move from his childhood into his teenage years without undue defiance? Maybe he just remained quietly analytical. I am curious about how he managed to deal with his youth.

He was only confined in the beginning. If he was to excel in school (college) and then medical school, he needed to break free from that confinement. He was very successful. His confinement was only that many of the traditions of his religion did not make sense to him, and he was released from those requirements early on. They were not an inhibitor in his development. He learns very fast. He can absorb information one time and then retain it forever. He would dismay his fellow medical students because he casually enjoyed free time, while the others studied intensely to absorb the same data. Such is his personality and skill.

Yes. His ability to easily absorb and retain information is a gift that he applied well.

Yes. He has seen countless human births and has treated them as patients over the years. One of Steve's step-relatives was his patient. Steve's niece was a patient of Doc. It is only coincidence. Considering his large practice, the probability was a higher one.

Did he know Steve at that time?

No.

May I ask you about Spencer?

Spessor?

Spencer. He was one of a couple that Doc, and Steve and I worked with on an abduction case.

Yes. And Leslie.

I am just wondering how he is doing since his operation since I do not hear from him.

He is healing slowly, slower than normal, due to his health issues and addiction to tobacco. It affects his ability to heal. This was known beforehand, but the alternative would be death, so this is the best outcome. His recover is slow, but it is successful. He will never be completely healthy. Leslie has been a good, attentive companion. Leslie is anxious. She often remembers the lights and the trauma of a cruel species who intruded on their vacation, yet there is strength between Spencer and Leslie. They find support that few will fully appreciate. They regard each other as survivors in life and in their strange experiences together. This has given them consolation. They have also appreciated the concerns and skills of you, Mary, Doc, and Steve, in assisting them in accepting the intrusions into their lives by another species that performs illegal actions. There are many abductions. Yet it is not realized that our species has prevented many more from happening, but it is a war between those who wish to trespass and those we try to block. Technology and skills are ever changing in this struggle. New means of traveling or intruding occur. This is countered, which is countered again by the opposing species. They wish to steal genetic material and other things that are all for their own selfish needs and to use for exchange on the black market with other species. Human genetic strains evolve quickly and are very adaptable, making them very desired by some other species for their own use. We try to minimize or stop this practice.

How ironic that there are human poachers of species here on Earth, and then I learn that there are some off-planet beings, alien cultures, who are poachers of human genetics.

Alien cultures?

Yes.

There are some off-planet beings that behave like human poachers. What comes around goes around.

Yes.

However, in this circle, some of the actors (humans) are unaware of the other actors. If they are aware, they may refuse to accept it. This is a way to exploit the situation. This is done by planting disbelief and denial in humans that this activity occurs or that such actors (abductors) even exist. This is used (human disbelief) to the advantage of the poachers.

I am puzzled why humans fail to see this activity for its true purpose.

It is your Earth centeredness.

I am puzzled as to why, with life existing in the Universe for so long, that there hasn't been much more maturity coming from those life forms, just by virtue of experiencing so many interactions with so many cultures. I would think that the experiences would cause species to develop full cooperation and compassion.

For the most part, this is true. You can take heart and solace in that. However, there are always a few renegades who, for whatever reason, feel that they are at a disadvantage. For their own need and purposes, there will be poaching to enhance what they feel are inadequacies in their species of genetic makeup. Humans are not the only species that are abducted. There are abductions on other planets of other species, as well.

From what I am understanding, you are saying that this is not normal behavior (abductions).

Correct.

Thank you. That is a relief.

This is not the norm. Just like in your society, not all are thieves, yet all are capable of theft. Their individual situations do not require them to take such risks. Mainly, in an accepted community, such behavior is inappropriate, yet there are always a few, for whatever reason, who will become selfish.

Thank you for that comparison.

It is an accurate comparison. The vast majority are not thieves.

You are familiar with the person known as Roger Lear, a Podiatrist?

Yes.

I think he does some surgery.

He is in general practice and specializing in Podiatry. He is a foot doctor and has been a pioneer in hoping to disclose the visits of alien contact by removing and examining alien implants.

He was scheduled for surgery. Just prior to the surgery he had an encounter and was healed. Would you comment on this?

His condition was treated so that he could survive surgery. Without this, he may not have survived, yet his encounter could have easily corrected his main affliction as well. However, this would be too obvious to everyone. We let human medicine take its course. Sometimes we provide assistance. They are usually unaware of a little assistance on our part.

Was he aware of the level of your assistance?

He is unaware, but we know he would be appreciative. Someday he may learn more.

The Universe

I came across a science program that showed a sea creature that was composed of many very small creatures. It looked rather like a jelly fish, but it had many small creatures attached together with a kind of single consciousness in making movements. Also, in a salt water lake in the

Palau Islands, there are jelly fish that carry algae in a kind of pouch. They tend the algae by taking themselves near the surface of the water to get sunlight for the algae and then rest near the bottom of the lake to take on nitrogen. In essence, they were growing the algae and then eating it.

It is a symbiotic relationship.

I was wondering if you have sea creatures that display farming behavior. Perhaps you have creatures that cluster together and work as one.

They form analogies of our society functions. Their body functions and their mission are to live successful, reproduce, and become a successful design for life. Other species on other planets have similar functions. Some planets have life forms that exist near higher elevations in the atmosphere. They absorb energy from the sunlight via biochemical means. Sometimes they will reproduce at lower levels. In the migrations of these strange species, they often become vulnerable to other predators and create an environment that benefits the other. This is much like your plankton. Many planets have very thick atmospheres that are not suitable for our forms of life, but they do exist in the atmosphere. They often have the ability to live and float in the atmosphere. They gather food and reproduce. At times they use the atmosphere to warm themselves to ascend or descend in the atmosphere, like living balloons. They will use the heat of the sun of their world to warm themselves and the natural gases that expand their bodies. This enables their bodies to become more buoyant, and they rise higher. By expelling the gas, they descend.

Like a hot air balloon.

They are living hot air balloons. There are other predatory species that will prey on these and use the energy that they absorb to nourish the predatory species. In turn, another species will use them in the same way. Such is the food cycle.

It sounds like something that was seen (during a NASA mission) in outer space may have been an escapee from one of these planets. It was an extremely large, mostly transparent, worm-shaped life form. I don't know that this life form could have sustained itself in deep space, where it was observed.

Able to migrate beyond its atmosphere? It is theoretically possible.

When a Zeta kin observes something that is completely outside of the experience of the observer or the entire species, what goes on in the mind to process this new information? How is this new information processed?

We follow the natural order of comparing to the past knowledge the new knowledge that has been introduced. We tend to categorize and classify into solid, liquid, gas, and life form type, if it is life. Also, the ability to recognize what constitutes life is a constant challenge. With new knowledge, we have discovered that life occurs in so many varied forms. Life abounds in the Universe. I have spoken of life in the form of energy consciousness. One species is a form of pure energy.

You had no frame of reference in your past history for this kind of life form?

Correct. And yet, we recognize it as life. It has consciousness. It creates waste and requires energy. It has all the qualifications for life in a form that may not seem to be easily recognized. Humans will learn to recognize unusual life forms. It is often so varied that one may not recognize it as life. This is a challenge for us, too, though, as scientists, we have a larger base of knowledge to make comparisons and classifications.

In that case, may I ask you a rather peculiar question that you may find amusing?

There are no peculiar questions in the context that we are having a peculiar conversation, and in that context, it is most welcome.

I am wondering about mathematics. When viewed as a kind of shorthand for handling very large and complicated quantities in conceptual arrangements and getting it to provide proof of concepts, it is the accepted tool for describing the Universe. Do you think that the universal elegance of mathematics indicates a kind of elegant consciousness behind it?

A consciousness can embrace mathematics, but mathematics cannot embrace consciousness, but in understanding the Universe, it is required that mathematics creates an understanding within that consciousness. There is great beauty in mathematics, and its commonality is a useful form for communication as well as enjoyment of many things, including music, which we have discussed. Mathematics is a universal of the Universe in explaining the reality of the Universe. It exists everywhere. It places a value and proportion as well as allowing us to penetrate larger mysteries of reality.

I don't know how difficult it is for you to communicate with this 'pure energy species' that you reference.

Our communication is not without difficulty. We realize they are not in some distant place. They are, but they are also as close as another of our species because they can exist anywhere in their dimension. It (communication) is a matter of us learning how to adapt to them to create a path of communication. Telepathy has been the prime means of contact so far, but it requires the highest level of discipline and focus. Many in our own species are unable to achieve this level, but we have, within our Zeta kin, a number of beings who have learned to specialize and adapt their thought processes to form those links of communication, but it has been a stretching time for us. It is worth the effort.

When I look at mathematics, I see it as a tool and process that is very probably used in other dimensions as well as this dimension. The fact that mathematics exists and is virtually everywhere makes me wonder about its initiating point. If you happen to come across some other sentient beings who have a different way of looking at the initiation of mathematics in all things, I would be interested in this subject.

Yes.

Other

When I was at a service this morning, I mentally invited you, Zestra, to attend because part of the service included doing healing work.

I was with you in thought and in touch.

Thank you. I wanted you to have the experience of how some human healers work.

We are healing sisters, across space, but sisters, never-the-less. We know the importance of healing, contentment and good health.

Yes.

Was the healing successful?

I believe it was somewhat successful.

I will add my thoughts to yours, if they will be of assistance. I will not interfere with your skills.

I believe any good thoughts are of assistance, and that good intention is the driver.

I say that to respect your skills.

Han has also become aware that good thoughts are of service. When I had a very minor illness, he sent his good thoughts to me. All of those good intentions are of service in speeding the healing process.

The energies are collected. You exploit them to their fullest.

I think these energies and the accompany intentions must attract each other.

Healing is like that, first a glimmer, then a flow, and then it compounds and blossoms into a full expression of healing that perpetuates itself.

As you are probably aware, I sent Steve off to collect species for his terra-forming project. Where is he at with that collection? (This was hypnotic suggestion given to Steve to occupy him while this therapist engaged with the Zetas.)

Yes. He is bringing any manner of species, from bacteria to the largest land animals and the largest of sea animals, especially fish, very exotic fish, much like a garden with koi. They will not only be aesthetically pleasing, but will fulfill an important ecological function.

He will miss his partners. They will not be gone. It is just that the tasks are approaching their completion. I am sure there will be regular maintenance on these projects in the future. Nicole and the others have been very special to him, as he has been to them. They are a strong unit. Thought and purpose were shared by them with great focus and often intense concentration to outdo themselves this time. The Planet Bonestall is complete. It is quite remarkable, and Jess Lee will be very pleased in spirit. He will run through the forest with his arms out like a child running through a new house, touching the walls and enjoying the new spaces with delight. This is what I see. He (Steve) sees two-dimensional concepts in three-dimensions, such as a painting, with another dimension added.

Yes. He is very visual. It is one of his gifts.

Steve named his new planet Bonestall after Chesley Bonestall. When Bonestall was a child, he went to the Lick Observatory in 1906 and saw Saturn through the telescope. It changed his paintings and his engineering illustrations. His being both an artist and an engineer makes the name of this planet all the more appropriate. He was a remarkable man, and he helped envision early space flight in his time. He was present in spirit at man's newest evolutionary step toward the stars. He showed the way. He was a San Francisco resident.

I hope his spirit is aware that Planet Bonestall was created to honor him.

He is aware. The young planet shows new growth, which will mature in time into a rich garden with a very dynamic and successful ecosystem. *(This project was initiated as a thought form in Steve's mind. Others from the Zeta society, joined in to work with Steve at the level of consciousness where thought is so detailed and complete that it is ready to manifest itself into the physical world. It could also be viewed as a state of consciousness that does not require physicality to be enjoyed.)*

I understand that your minds can create very complex and realistic settings, which you can enjoy. I think Steve has the kind of mind to create these very detailed environments in this way. I believe this has been successful.

This has been a good lesson for Steve in expressing his skills. Also, the lesson has been that he has the skills. He has only been confined by physical means, but his mental means, in concert with his colleagues, combined with receiving our thought in teaching him how to conceptualize. It has been a rich experience for all.

Open Session for Comments and Questions

I just wanted to say to Zestra that I appreciate your being present with me today when I was at a gathering, doing healing work.

Zestra is present, but, unusually quiet. Her thoughts are merged with mine. I am primarily speaking to you today, but my responses include both of our thoughts. Zestra is just as imprinted in these responses as I am.

You may both be aware that I am working on and following George's methodology for remote viewing.

It will be a successful path.

It will be a good stretch, and I do not mind stretching. I hope I can find good and useful ways to apply it.

It will mainly be opening new avenues for communication. You can anticipate this. This will come with your 'new found country,' to quote Shakespeare.

That is a good way to describe it.

It will open up new worlds to other species.

Do you have any suggestions for Steve in giving him opportunities to expand and experience new abilities to throw switches so that he can reach out more and enjoy the complexity of life around him?

He appreciates that. He would embrace new concepts. He is being very effective now by having the confidence to accept that he could pursue those abilities. Recognizing his ability is as important as the willingness to explore it. If he is made aware, he is naturally curious and, with your guidance, Steve will enjoy much more exploration. This has been a credit to you. He appreciates that, and we understand and appreciate you, too.

Thank you. I don't see why people put so many boundaries on themselves. Maybe it is just society that causes it.

He has been tricked into conforming to the average, when he is not.

He is much smarter than he has permitted society to know.

He is. He is in a deep place, now, in his mind. Sometimes it is hard, at this moment. In a way, he is very open to listening to us, but he is struggling to stay focused in his connection with you. He is in a dark place now, but he is safe.

It is time to bring in some warmth and light to him.

He is in deep water.

I don't like to part.

We are not parting. We are together always.

Yes.

These moments are very special, and they are important. It is part of a special relationship we share and enjoy. Our visits will become less frequent only because you need to focus and prioritize your time, but know that we are always available for you whenever you wish. Because the task you are doing is very important, we honor and respect it. You are going to share our thoughts with many others of your Earth kin, and this will give us happiness. We wish we could answer the inevitable follow-up questions, but we will be available to you for that purpose when the appropriate time comes.

Most of what you have given to me will be released (published).

In some ways, this is unprecedented, but it is time to widen the doorway and build a new doorway with more layers of communication. This is the need for disclosure, not a political one, but one that is between being and being, Zeta kin and Earth kin, and we are well pleased. We are not disclosing anything that will compromise, but simply what is necessary at this time. It is good.

I feel quite good about it, too.

As do we, and we know by your culture, our names could not be in the book as credit, but there are many authors to this book. But you and Steve will be the primary ones, with you doing the writing, and Steve translating his images into visual form for the enjoyment in the book. You will be our ambassadors.

It is good to do work that is of service to others. Zestra, I wish you a fond adieu and Namaste.

Author's Notes:

These contacts also occurred between the Zetas and me, using the vehicle of Dr. Gene Lipson, a retired pediatrician. They enjoyed coming in through 'Doc' because they found his belief systems interesting, although not necessarily accurate. It is hoped that these contacts will continue for years to come. It remains to be seen because Steve, Doc, and I are mortal and in our later years. I have felt compelled to share this astounding information because humanity has a right to know. Through this process I have become acquainted with others who receive direct communication from the Zetas. One exceptional contact is William Treurniet and the medium he works with named Paul Hamden. We are only two of hundreds of contacts, according to the Zetas. The Zetas are willing to communicate with those who can hear them without immediate judgment, and who are open to examining and expanding our belief systems. Many Zetas see humans as violent and not worth the effort of communication. The Zetas Han, Zestra, and Gen also work with their own species to enable them to take a gentle view, perhaps seeing mankind as worth the effort.

Dear Reader: Since you have concluded reading this book, you have encountered many new concepts from the Zetas. Please share this surprising and mind-expanding body of knowledge with those who are willing to learn.

www.ingramcontent.com/pod-product-compliance
Lightning Source LLC
Chambersburg PA
CBHW031832090426
42741CB00005B/214